GET A FIRMER GRIP ON YOUR MATH

William J. Adams

with Illustrations by

Ramuné B. Adams

KENDALL/HUNT PUBLISHING COMPANY
4050 Westmark Drive   Dubuque, Iowa 52002

To RASA ADAMS

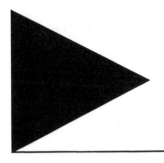

CONTENTS

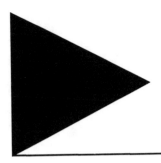

PREFACE

This sequel to *Get a Grip on Your Math* is intended for readers who wish to dig deeper. It provides food-for-thought questions and a more in-depth discussion of some of the basic ideas developed in *Get a Grip on Your Math*. This includes discussion of progressions and the mathematics of money, further discussion of sampling, validity versus truth, linear programming extensions of Mathematics for Business and Economics, and probability model extensions of Mathematics and Chance.

Taken together, these books are suitable as texts for liberal arts, education and other majors in courses which go under such headings as *The Development of Mathematics*, *Mathematics for Liberal Arts*, *Mathematics and Its Applications*, and *The Nature and Structure of Mathematics*. They are also suitable for people who have been diagnosed as suffering from "math anxiety."

W.J.A.

R.B.A.

ZERO, MILLIONS, BILLIONS, AND MORE

1.1 ZERO

"Did you know that 0/0 is 5?" John asked his friends Jim and Burt. "It follows from the following mathematical analysis," he exclaimed.

"Consider the ratio:

1

$$y = \frac{x^2 + x - 6}{x^2 - 3x + 2}$$

For x = 2, calculation yields y = 0/0. But the numerator and denominator of the preceding ratio can be simplified by factoring. This gives us:

$$y = \frac{(x - 2)\,(x + 3)}{(x - 2)\,(x - 1)}$$

Cancellation of (x − 2) from the numerator and denominator of the preceding allows us to write y in the simpler form:

$$y = \frac{x + 3}{x - 1}$$

For x = 2, calculation yields y = 5.

Since y = 0/0 and 5 for x = 2, it follows that 0/0 = 5."

Food for Thought

1. Do you agree with John's analysis? Explain.

2. Do you agree with Jim's conclusion that 0/0 = 1 and his reasoning? Explain.

3. Do you agree with Burt's conclusion that 0/0 = 0 and his reasoning? Explain.

4. Do you agree with Burt's reasoning in rejecting John's conclusion that "what this means is that 5, 0, and 1 are equal?"

1.2 MILLIONS, BILLIONS, AND MORE

Food for Thought

1. Colleges and universities have received considerable donations from generous donors in recent years. Translate the following donations into equivalent time.

    a. Harvard Medical School; Isabelle and Leonard Goldenson, $60 million, 1994.

    b. Emory University; Robert Woodruff, $105 million, 1992.

    c. University of Pennsylvania; University of Southern California; $120 million, Walter H. Annenberg, 1993.

    d. New York University; between $100 million and $500 million in art and property, and $25 million in endowment, 1994.

2. In May of 1994 General Motors announced that it would reduce its pension fund deficit by adding $10 billion to its maintenance.

a. Translate this amount into equivalent time.

b. What time period does this take us back to and what were some of the developments in this period.

3. In June of 1994 Governor Mario Cuomo of New York and legislative leaders announced that they had reached agreement on a $34 billion state budget. Same questions as 2(a) and 2(b).

4. Examination of President Clinton's spending package proposed in April of 1993 shows the following estimates for 1994 (in billions of dollars).

    i.   NASA: 14.7

    ii. Housing and Urban Development: 28.9

    iii. Transportation: 39.1

    iv. Agriculture: 63.0

    v.   Defense: 264.2

    vi. Health and Human Services: 640.1

    Same questions as 2(a) and 2(b).

5. Buck Rogers, as the story goes, was frozen in an accident which occurred in the 20th century and reawakened 500 years later. Buck's first action was to call his broker, who, of course, was not available. The firm was still in existence and Buck was pleased to learn that his investment of years before was now worth $8.5 million. His delight with the news was tempered, however, when the operator told him to deposit $50 for the next 5 minutes. The state budget, Buck subsequently learned, was $100 trillion and was expected to rise to $126 trillion within 5 years.

    Same questions as 2(a) and 2(b) for the $100 trillion and $126 trillion figures.

6. The 1990 census determined the population of the United States as 248.7 million. Translate this number of people into equivalent time (1 person = 1 second) in years.

7. As part of a coding system a 129 digit number termed R.S.A. 129 was put forward by Drs. Ronald Rivest, Adi Shamir, and Leonard Adleman in 1977 with the challenge to break it into its component parts. With the methods available at the time it was expected that the problem would stand well into the 21st century. The problem was solved in April of 1994. It required 100 quadrillion calculations (1 quadrillion = 1000 trillion). Translate this number of calculations into equivalent time (1 calculation = 1 second) in years, centuries, and millennia.

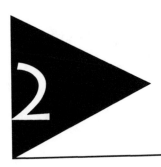

2 SLIPPERY FIGURES AND TALOSIAN IMAGES

2.1 FOOD FOR THOUGHT

In replying to questions 1–26 consider the articles cited to establish a context and background, and other relevant information that you might obtain.

1. What caused the asbestos panic in New York? S. Dillon, "New York City School Asbestos Tests are Voided," *The New York Times*, Aug. 7, 1993.

2. Are "light" cigarettes really light? P. Hilts, "Major Flaw Cited in Cigarette Data," *The New York Times*, May 2, 1994, A1.

3. How did the rats upset the data cart? J. Bishop, "They Smell a Rat and it Really Raises their Blood Pressure," *The Wall Street Journal*, June 10, 1994, A1.

4. Why is the Congressional Office of Technology Assessment disturbed by the Congressional Budget Office's analysis of competing health care proposals? R. Pear, "Second Thoughts on Health Data," *The New York Times*, May 9, 1994.

5. Who's upset with Chinese statistics and why? S. Tefft, "China is Under Pressure to Clean Up its Statistics," *The Christian Science Monitor*, June 9, 1994, 8.

6. Do the data show the true state of violence in New York Schools? S. Dillon, "Violence Grows in New York Schools," *The New York Times*, June 11, 1994, 25; V. Toy, "Draft Audit Says Board of Education Understates Crime in Schools," *The New York Times*, Sept. 2, 1995, 24.

7. Were educational gains as good as they were claimed to be? W. Celis 3rd, "Group Admits School Gains Weren't Real," *The New York Times*, June 7, 1994, A21.

8. Are these figures fact or fiction? P. Montgomery, "Playing with Numbers," *Common Cause Magazine*, Sept/Oct 1987, 38–39.

9. Has the tobacco industry been blowing smoke on secondary smoke? P. Hilts, "Data on Secondhand Smoke Were Faked, Workers Say," *The New York Times*, Dec. 21, 1994, D23.

10. Were data withheld on jet engines? "G.E. Denies it Withheld Data on Jet Engines," *The New York Times*, June 6, 1994, D3; D. Frantz, S. Nasar, "F.B.I. Inquiry on Jet Engines New Jolt to Company Image," *The New York Times*, July 18, 1994, A1.

11. Are Britain's unemployment data reliable? "Measuring Up," *The Economist*, Jan. 23, 1993, 58–59.

12. Has the nation's system of gathering accurate statistics broken down? M. Lewis, "Sloppy Statistics, Poor Policies," *Journal of Commerce and Commercial*, April 2, 1993, 6.

13. Does the government's economic statistics differ from reality? P.S. Nadler, "More Fantasy Figures from Your Uncle Sam," *American Banker*, June 1993.

14. How were tests on a missile system's component faked? J. Rabinovitz, "Company Admits it Faked Tests on Missile System Part," *The New York Times*, May 17, 1994.

15. It has been argued that the GNP has been inflated because of a numbers game. What is the numbers game and in what sense has the GNP been inflated? A. Zipser, "Numbers Game: Why GNP Has Been Consistently Inflated," *Barrons*, Oct. 15, 1990.

16. It has been argued that payroll statistics and household data are partially relevant to describing economic growth. What is the basis for this point of view? G. Korentz, "The Payroll Numbers May Be Hiding Economic Trouble," *Business Week*, Nov. 20, 1989.

17. Why did Synergen halt trials with its "flagship" drug Antril? Does Synergen's experience have any lessons for us? L. Fisher, "Sepsis Drug Trials Halted; Stock Dives," *The New York Times*, July 19, 1994, D1.

18. Was the testing of New York City's drinking water compromised? A. Finder, "New York City Official Says Tests of Water Were Skewed." *The New York Times*, Jan. 13, 1995, A1; D. Van Natta, Jr. "Albany Decides New York City's Water Tests for Bacteria Were Thorough," *The New York Times*, Aug. 10, 1995, B4.

19. The Crime Bill passed by Congress in August 1994 provides for adding 100,000 cops to the nation's streets. But is there less here than meets the eye? T. Gest, G. Witkin, "A Cold-Eyed Look at Crime," *U.S. News & World Report*, June 27, 1994, 33–34.

20. Is the U.S. trade balance for 1991 –$28 billion, $24 billion, or some figure in between? "A New Look at Trade," *Fortune*, Aug. 22, 1994; 130.

21. In what sense is Medicare inpatient operating margin data misleading? L. Wagner, "Medicare Data Misleading—ProPac Chief, *Modern Healthcare*, April 24, 1994; 2,10.

22. Every year the New York State Health Department issues a ranking in terms of patient mortality rates for surgeons who perform coronary bypass surgery in the state. But is a surgeon's ranking enough to indicate his competence? E. Bumiller, "Death-Rate Rankings Shake New York Cardiac Surgeons," *The New York Times*, Sept. 6, 1995, A1.

23. Is it common practice for computer companies to skew data on the speed of microprocessor chips? J. Markoff, "Intel Admits it Overstated Chips' Speed," *The New York Times,* Jan. 6, 1996; 33.

24. The 1995–96 push and pull match between the Dole-Gingrich camp and the White House suggests consideration of the following questions:

   a. The sound of budget-in-balance has a captivating ring to it, but is this numerical relationship the one we should focus on in connection with the size of the nation's gross national debt and its impact on the economy?

   b. Is the term budget-in-balance unequivocal or is it possible for two balanced budget advocates to be talking about very different budgets?

c. After a tug-of-war on number credibility, it was agreed that the numbers and projections of the Congressional Budget Office should be used as a basis for budget balancing negotiations. But are these numbers and projections reliable?

W. J. Adams, *Get a Grip on Your Math* (Ch. 7, Sec. 4); R. Eisner, "We Don't Need Balanced Budgets," *The Wall Street Journal*, Jan. 11, 1995; P. Grier, "Behind the Budget Bluster," *The Christian Science Monitor*, Dec. 19, 1995, p. 1; P. Passell, "Economic Scene," *The New York Times*, Dec. 21, 1995, D2; L. Uchitelle, "Politicians May Be Up in Arms About Government Deficits, But Economists Aren't," *The New York Times*, Jan. 8, 1996, p. 13.

25. What financial protection does the Federal Government have against research fraud? P. Hilts, "2 Universities to pay U.S. $1.6 million in Research Fraud Case," *The New York Times*, July 23, 1994, A9.

26. Locate two articles which take issue with data from the point of view of underlying assumptions, reliability or relevance. Write a commentary for each article explaining their authors' points of view. Possible sources for such articles include newspapers and magazines which address serious domestic and international issues, such as *The New York Times, Christian Science Monitor, Wall Street Journal, Newsweek, Business Week, Time,* and *U.S. News & World Report,* as well as professional journals concerned with your area of professional interest.

27. Locate two articles whose arguments are based on figures and subject them to the test questions developed in Chapter 9 of *Get a Grip on Your Math*. What do you conclude?

28. a. What is the difference between nominal and ordinal data?

    b. What is the difference between interval and ratio data?

29. Classify the following examples of data as nominal, ordinal, interval, or ratio. Explain the reason for your classification in each case. (a) Eye color, (b) Religious affiliation, (c) Course grade (A, B, C, D, F), (d) IQ score, (e) Height, (f) Weight, (g) Ranking of twelve football teams in a citywide competition, (h) Temperature in the Fahrenheit scale.

WHAT DO THE NUMBERS SAY?

3.1 IT DEPENDS ON HOW YOU LOOK AT IT

"I like numbers because they're precise," said Bottom-line Bob to Reflective Ramunė, "Give me the numbers and that ends all arguments; no more beating around the bush." "Does this mean there's no room in your number world for others who might seek to make sense of the numbers?" inquired Ramunė. "What's there to make sense of?" replied Bottom-line Bob. "Numbers speak for themselves." "But through whose judgment, and how reliable is that judgment?" countered Ramunė. "My judgment," replied Bob; "that's what Bottom-line means." "Yes, but I'm afraid that may also be quicksand masquerading as bedrock," answered Ramunė.

Food for Thought

1. Ellen Ames and Ann O'Neil, professors of economics at Huxley College, saw a grade of 50 on Professor Ames's last economics exam in different terms. "A grade of 50 is an F," noted Professor Ames. "But it's the highest grade in the class," replied Professor O'Neil, "and as such an F does not make sense." Discuss the significance of 50 as the highest grade on an exam, and as an F.

2. "Herman is a great pitcher because he has an earned run average of 2.54," Luis pointed out to his friend Jim. "Herman is mediocre because he has a 125–120 won-loss record," replied Jim to Luis. Who's right? Discuss.

3. The larger the police force of a community, the smaller the crime rate, right? (R. Moran, "More Police, Less Crime, Right? Wrong." *The New York Times*, Feb. 27, 1995, A15; D. Francis, "Want Less Crime? Hire More Cops," *The Christian Science Monitor*, March 3, 1995, 9).

4. One thousand senior high school students in Ralph City took a college level math course, with 80% of them passing the state-wide standard exam. The following year 3000 senior high school students in Ralph City took the course, with 50% of them passing the statewide standard exam. Do these figures give us cause for pessimism or optimism concerning Ralph City's education system?

5. "80 percent of the crimes in this city were committed by Plutonians," remarked Oscar to his wife Janet. "They're a bad lot." Are there other possible interpretations of this figure? Are there other figures that might be relevant? Discuss.

6. "80 percent of the A grades in my math class were earned by Plutonians," commented Bob Levy to his colleague Amy Flores. "This suggests that they have an innate ability for math." Are there other possible interpretations of this figure? Are there other figures that might be relevant? Discuss.

7. "The proportion of black and Hispanic students in New York City public schools who read and do math at the level expected for their grade is far below that of white and Asian students, new test

results show." C. Jones, "Test Scores Show Gaps by Ethnicity," *The New York Times*, July 8, 1994, B1; A. B. Jeffries, "Schools Can't Give Students What They Don't Get at Home," *The New York Times*, July 20, 1994, A18.

    a. What are some possible explanations for these test results?

    b. What reforms do the explanations suggest?

8. Is it possible for a best selling book not to make it onto a best-seller list such as *The New York Times* or *Publishers Weekly* lists? (S. Lyall, "Publishers' Aces in the Hole: Books that Sell Well Quietly," *The New York Times*, Aug. 1, 1994, A1.) Is it possible for a non-best seller to make it onto a best-seller list? (A. Marks, "Bestseller Lists Bind Industry in Controversy," *The Christian Science Monitor*, Sept. 6, 1995, 1.)

9. "Urban Crime Rates Falling this Year" (C. Krauss, *The New York Times*, Nov. 8, 1994; p. A14) strikes an encouraging note in an era dominated by dismal crime statistics. But are crime rates really falling? (Also see, for example, "Shining a Light on Hidden Crime," *U.S. News & World Report*, Nov. 7, 1994; p. 8.)

10. "Blacks Prone to Dismissal By the U.S.: Their Discharge Rate Tops Other Workers," (K. De Witt, *The New York Times*, April 20, 1995, A19). Does this mean that blacks have been discriminated against when it comes to dismissal? Discuss.

11. Only 6 percent of the people quoted on page 1 of *The New York Times* during February, 1990 were female, observed Betty Friedan in a survey by her group *Media Watch*. Is this an insightful statistic or misleading figure? (J. Leo, "No, Don't Give me a Number!," *U.S. News & World Report*, May 14, 1990, 22.)

12. Medicare, which provides health care for about 37 million Americans, is about to go broke, we have been told by Democrats and Republicans. Assertion: The Republicans intend to cut Medicare. Reply: Republicans do not intend to cut Medicare, just slow the rate of growth in spending. What do the figures say? Which figures are most meaningful and from what point of view? (See, for example, D. Rosenbaum, "The Medicare Brawl: Finger-

Pointing, Hyperbole and the Facts Behind Them," *The New York Times*, Oct. 1, 1995, 18.

13. Locate two situations for which conclusions are presented based on figures cited. Are there other interpretations for the figures cited? Are there other figures that might be relevant? Discuss.

3.2 GET THE FIGURES

"Let's get the figures" cried out the observer, doggedly pursued by reality. Sound advice, when the figures are sound.

Food for Thought

1. One stereotype of the behind bars population is that the "typical" prison inmate is a black jobless high school drop out from a broken home. But what do the figures say? (See, for example, R. Morin, "Redrawing the Face of Crime," *The Washington Post National Weekly Edition*, Oct. 10–16, 1994; p. 50.)

2. In recent years the view that electromagnetic fields generated by power lines is a cause of cancer has led to a contentious public health issue. Are electromagnetic fields generated by power lines as much a concern as the earth's magnetic field and the magnetic fields generated by home appliances? What do the numbers say? (See, for example, W. Broad, "Cancer Fear is Unfounded, Physicists Say," *The New York Times*, May 14, 1995, A19.)

3. Three strikes and you're out, or three strikes and you're broke? "Three strikes and you're out" has become an anti-crime war cry and many "crime fighters" have joined the crusade. Crusades cost and the question is: What are the costs? (See, for example, W. Claiborne, "Three Strikes Laws Have a Price," *The Washington Post National Weekly Edition*, August 15–21, 1994; p. 33; J. Gray, "New Jersey Senate Approves Bill to Jail 3-Time Criminals for Life," *The New York Times*, May 13, 1994; p. A1.)

4. The image of sex in America communicated by much of the popular media is one dominated by extramarital affairs and rampant casual sex.

    a. What do the figures say? Note references cited below.

    b. An important related question is: How reliable are the figures? (See Section 5.1, Question 18.)

PRIMARY SOURCES

1. E. Laumann, J. Gagnon, R. Michael, S. Michaels. *The Social Organization of Sexuality.* (Chicago: University of Chicago Press, 1994).

2. R. Michael, J. Gagnon, E. Laumann, G. Kolata. *Sex in America* (Boston: Little Brown & Co., 1994).

SECONDARY SOURCES

3. T. Lewin, "Sex in America: Faithfulness in America Thrives After All," *The New York Times*, Oct 7, 1994; p. A1.

4. J. Schrof, B. Wagner. "Sex in America," *U.S. News & World Report*, Oct. 17, 1994; pp. 74–81.

5. Asbestos has been treated as a public health menace and expensive clean-up efforts which could reach as much as $30 billion in cost have been undertaken. The question is: In terms of the health danger posed by asbestos, does the risk to health warrant the cost of the clean-up campaign? Are there "higher" health priorities? What do the figures say? (See, for example, P. Cary, "The Asbestos Panic Attack," *U.S. News & World Report*, Feb. 20, 1995, pp. 61–63.

6. According to free-market theory lowering wages leads to higher employment. What do the figures say and what is their significance? See: W. Hutton, "Minimum Wage Offers Maximum Returns," *Manchester Guardian Weekly*, July 23, 1995; 21.

7. Take up two situations (defense spending, health care, space exploration, etc.) from one point of view or another, which have cost implications. Explore the cost implications, noting the underlying assumptions. Do the cost implications call into question the viability of the proposal or scheme?

8. Every year *U.S. News & World Report* ranks America's best colleges and universities. In its September 18, 1995 issue teaching was ranked for the first time (Head of the Class, p. 140). This raises questions of what factors should be taken into account and how they should be taken into account in developing a numerical measure of teaching effectiveness for institutions and individuals. Discuss this issue.

3.3 GEOMETRIC PROGRESSIONS

As we saw in "Silence May Be Golden After All" (Ch. 5 of *Get a Grip on Your Math*), Senator Lloyd L. Wind was able to turn silence to gold because of his knowledge of geometric progressions. They have a number of interesting and important applications which we examine more fully in this and the next section.

A sequence of numbers in which each term is obtained by multiplying the preceding term by a constant r is called a **geometric progression**. The constant r is called the **common ratio**. If a is the first term of a geometric progression with common ratio r, the first n terms are:

$$a, ar, ar^2, ar^3, \ldots, ar^{n-1} \tag{1}$$

Thus, for example, the first four terms of the geometric progression whose first term is 2 with common ratio 3 are 2, 6, 18, and 54.

Let S_n denote the sum of the first n terms of the geometric progression described by:

$$S_n = a + ar + ar^2 + ar^3 + \ldots + ar^{n-1} \tag{2}$$

Multiplying by r yields:

$$rS_n = ar + ar^2 + ar^3 + \ldots + ar^{n-1} + ar^n \qquad (3)$$

Subtracting (3) from (2) gives us $S_n - rS_n$ on the left side and $a - ar^n$ on the right side, since the ar, ar^2, ..., ar^{n-1} terms subtract out.

Thus, we have:

$$S_n - rS_n = a - ar^n$$

Factoring yields:

$$S_n (1 - r) = a (1 - r^n)$$

Dividing both sides of this equation by $(1 - r)$ gives us:

$$S_n = \frac{a (1 - r^n)}{1 - r} \qquad (4)$$

as the sum of the first n terms of a geometric progression with first term a and common ratio $r \neq 1$.

Example 1

Find the sum of the first 6 terms of the geometric progression whose first term is 1 with common ratio 2.

Here the role of a is played by 1, $r = 2$, and $n = 6$. From (4) we have:

$$S = \frac{1(1 - 2^6)}{1 - 2}$$

$$= \frac{1 - 2^6}{-1}$$

$$= 2^6 - 1$$

$$= 64 - 1$$

$$= 63$$

Example 2 Beware the Wind

The common ratio r = 2 describes the procedure of doubling each term of a geometric progression to obtain the next term. Sums of geometric progressions increase modestly at first, but then take off. This was clear to Senator Lloyd L. Wind in his dealings with his colleagues who wanted to keep him quiet for a while. The agreement was that "beginning next Monday we will donate one dollar to the charity you designate and double the size of the donation every week thereafter for as many weeks as you remain silent in the Senate" (*Get a Grip on Your Math*, Ch. 5, p. 79). Wind stayed the course for 20 weeks, which generated a total of 21 payments, each payment being twice that of its predecessor. The sum of this geometric progression, with $a = 1$, r = 2 and n = 21, is

$$S_{21} = \frac{1(1 - 2^{21})}{1 - 2}$$

$$= 2^{21} - 1$$

$$= 2{,}097{,}151,$$

which required Wind's colleagues to pay his favorite charity $2,097,151.

Wind could easily had stayed the course another week, which would have payment given by:

$$S_{22} = 2^{22} - 1$$
$$= \$4,194,303$$

Staying silent an additional week would have brought payment of $8,388,607. But Senator Wind, while longwinded, was also considerate of his colleagues. He wanted to teach them a lesson for being "so smart," but one that they could afford.

Food for Thought

1. Find the sum of the terms of a geometric progression defined by the values of a, r and n.

   a. $a = 2, r = 2, n = 8$

   b. $a = 1, r = 3, n = 10$

   c. $a = 3, r = 4, n = 7$

   d. $a = 9, r = \frac{1}{3}, n = 12$.

2. Over a certain time period the number of bacteria in a culture doubles every minute. If there are initially 100 bacteria in the culture, how many will there be at the end of 30 minutes?

3. **Do you have enough funds for this?** Amy Allen, teenage daughter of Fred and Gail Allen, consistently practiced one basic principle of life: shop 'till you drop. Her parents believed that it was long overdue for Amy to come to grips with the "realities" of money and institute a savings plan. To help encourage her to get this going Fred offered Amy the following deal. "If you deposit $1 into your account each week, then I will deposit $1 into your account for your first deposit and double the size of my deposit every week thereafter for a year, provided that you maintain your deposits without interruption. If you miss a deposit, you

forfeit the entire amount that I will have deposited to that point."
If Amy agrees to this arrangement, how much will her father be
committed to depositing into her account?

4. **Reward Thy Servant.** Legend has it that Prince John of the Sea
    served his master Emperor Ronald the Astrologer faithfully at
    great risk to himself for many, many years. Finding himself an
    old man with not much to show for his years of devoted service,
    John requested an audience with the emperor.

    The emperor was delighted to see him and offered to secure for
    him a place in the Senate where many before had made their
    fortunes. But this was not what John wanted. He wanted a grant
    that would allow him to purchase an estate and live out his
    remaining years in comfort. "What sum would you consider
    sufficient for your needs," asked the emperor. "Ten million
    dollars," answered John of the Sea.

"I want you to receive your just reward," replied the emperor, "and this is what I suggest. Go to my Keeper of the Coin, who will give you one coin of silver weighing 1 ounce; bring it here. Go to him tomorrow and he will give you another coin worth twice the first; bring it here and place it beside the first. On the day after tomorrow he will give you a coin worth four times the first, on the day after that he will give you one worth eight times the first, and so on. Bring them here. You must do this yourself. When you can no longer lift the coin our agreement will have ended, but you may keep all the coins you will have taken from the Keeper of the Coin; that will be your reward." John of the Sea was delighted, for he imagined riches beyond his wildest expectations. He thanked Ronald the Astrologer profusely for his generosity and set about his task.

But John no longer had the strength of his youth and, although driven by desire, the best he could manage was a coin weighing 512 pounds.

a. How many coins will John take from the Keeper of the Coin?

b. If silver is worth $10 an ounce, what is the value of John's reward?

c. How does John's reward compare with his request for $10 million?

3.4 THE MATHEMATICS OF MONEY

MONEY GROWTH BASICS

In the world of finance **interest** is simply a fee charged for the use of borrowed money. It is an amount which is stated in terms of some monetary unit (dollars, cents, pounds, francs, etc.) Thus, if Mr. Debtor borrows $100 from Mr. Creditor with the understanding that $110 is to be repaid one month later, the interest charged Mr. Debtor is $10. The time period at the end of which the amount borrowed, called the **principal**, and the interest owed are to be paid is called the **interest period**. In the Debtor–Creditor transaction the principal is $100 and the interest period is one month. The sum of the principal and interest to be paid at the end of the interest period is called the **amount**. In the Debtor–Creditor transaction the amount is $110.

The ratio of the interest charged during the interest period to the amount of money owed at the beginning of the interest period is called the **interest rate**. The interest rate in the Debtor–Creditor transaction is:

$$\frac{10}{100} = 0.10 \text{ or } 10\% \text{ per month}$$

Although it is incorrect to do so, in colloquial language the terms interest and interest rate are used interchangeably.

More generally, let us observe that if P is the principal borrowed and I is the interest to be paid for the underlying interest period, the interest rate i is defined by:

$$i = \frac{I}{P}$$

The interest I for the interest period is given by:

$$I = Pi$$

(Interest) = (Principal) x (rate for the period)

If Mr. Debtor sought to extend his loan at the end of the month when it came due, Mr. Creditor might well argue that in extending the loan another month it would be appropriate to charge interest on the $110 now owed. Ten percent interest on $110 is $11 so that the total interest for the two months would be $10 + $11 = $21. The total amount would be $100 + $21 = $121.

In general, if the interest due is added to the principal at stated intervals of time and itself earns interest thereafter, the sum by which the original principal has been increased at the end of any time is called **compound interest**. The time interval between successive additions of interest to principal is called the **interest period** or **conversion period**. At the end of each conversion period the new principal, consisting of the original principal plus the compound interest, is called the **compound amount**. In the Debtor–Creditor situation, considered over two months, the compound interest is $21, the interest or conversion period is a month, and the compound amount is $121.

To obtain an equation for the compound amount let us suppose that an initial principal P is invested at compound interest at a rate i per interest period. Then at the end of the first interest period the compound amount is:

$$A_1 = P + Pi = P(1 + i)$$

At the end of the second interest period the new amount A_2 is A_1 plus the interest A_1i obtained from A_1.

$$A_2 = A_1 + A_1i = A_1(1 + i) = P(1 + i)(1 + i)$$
$$A_2 = P(1 + i)^2$$

At the end of the third interest period the new amount A_3 is A_2 plus the interest A_2i obtained from A_2.

$$A_3 = A_2 + A_2i = A_2(1 + i) + P(1 + i)^2(1 + i)$$
$$A_3 = P(1 + i)^3$$

More generally, at the end of the nth interest period the new amount A_n is given by

$$A_n = P(1 + i)^n, \tag{1}$$

where P is the principal initially invested, i is the interest rate for the interest period, and n is the number of interest periods.

Example 1

$1000 is to be invested 6 months from now at the rate of 3% per a 6 month period. What is the amount on deposit 24 months from now?

The interest period is 6 months, $P = \$1000$, $i = 0.03$, and $n = 3$. The problem is to find A_3. We have:

$$A_3 = 1000(1.03)^3 = 1000(1.09273)$$
$$A_3 = \$1092.73$$

Values of $(1 + i)^n$ are given in Table 3.1 at the end of this chapter, from which we obtain $(1.03)^3 = 1.09273$.

In the world of finance an interest rate cited as 6 percent per annum compounded twice a year (semi-annually) means that 3 percent interest is added every 6 months to the amount accumulated. More generally, the rate **r per annum compounded m times a year** envisions the year divided into m interest periods of equal length, with interest at the rate $i = r/m$ being added to the amount accumulated at the end of each period.

Replacing i by r/m in (1) yields

$$A = P\left(1 + \frac{r}{m}\right)^n$$

as the compound amount after n interest periods. To express n in terms of years let us first note that the number of interest periods in 1 year is m; thus the number of interest periods n in x years, let us say, is $n = mx$. Thus

$$A = P\left(1 + \frac{r}{m}\right)^{mx} \tag{2}$$

expresses the **compound amount after x years** when principal P is invested at the rate r per annum compounded m times a year. x can take on nonnegative integer values (x = 0, 1, 2, etc.) and fractional values consistent with m (x = 1/m, 2/m, etc.)

Example 2

Arthur Bryan plans to invest $1000 at 8% per annum compounded quarterly. What will he have on deposit in 5 years?

The problem, of course, is to recognize what we want and what we have. We want A. We have: P = $1000, r = 0.08, m = 4, so that r/m = 0.02; also x = 5, so that mx = 20. From (2) we obtain:

$$A = 1000(1.02)^{20} = 1000(1.48595)$$
$$A = \$1485.95$$

The situation considered so far involves investing a certain amount P, NOW, and determining what it's worth, A, LATER. Situations also arise where we envision the need for a certain amount of money A, LATER, and wish to find the principal P that we should invest NOW to obtain A, LATER. For instance:

Example 3

Piedmont Car Service envisions a need for $12,000 in 4 years to replace one of its cars. What principal should be invested at 9% per annum compounded 3 times a year if $12,000 is to be available in 3 years.

The problem is to find P. We have A = $12,000; r = 0.09, m = 3, so that r/m = 0.03. We have x = 3, so that mx = 9. Thus:

$$12,000 = P(1.03)^9$$

Dividing both sides by $(1.03)^9$ gives us:

$$P = \frac{12,000}{(1.03)^9}$$

Let us recall that negative exponents are defined by $a^{-n} = 1/a^n$. Thus, for example, $2^{-3} = 1/2^3 = 1/8$. In this situation we have:

$$\frac{1}{(1.03)^9} = (1.03)^{-9}$$

Thus, we may express P in the form:

$$P = 12,000(1.03)^{-9}$$

From Table 3.2 at the end of this chapter we have:

$$(1.03)^{-9} = 0.76642$$

Thus:

$$P = 12,000(0.76642)$$
$$P = \$9197.04$$

Piedmont Car Service should invest \$9197.04 now at 9% per annum compounded 3 times a year to obtain \$12,000 for a car replacement in 3 years. \$9197.04 is called the present value of \$12,000 with respect to the cited conditions. It expresses the present worth of \$12,000 to be realized 3 years from now.

More generally, solving

$$A = P\left(1 + \frac{r}{m}\right)^{mx} \tag{3}$$

for P by dividing both sides of (3) by $(1 + r/m)^{mx}$ yields:

$$P = \frac{A}{\left(1 + \dfrac{r}{m}\right)^{mx}} \tag{4}$$

In this setting P is called the **present value** of A. $P is the sum of money that must be invested now at the rate r per annum compounded m times a year if $A is to be realized x years from now. $P is what $A, x years in the future, is worth today under the cited interest conditions.

Food for Thought

1. Helen Needy borrowed $500 from Paula Plenty. The loan was for one year and the interest was $50.

    a. What is the interest rate?

    b. When the loan came due Helen asked Paula to extend the loan for a year at the same rate. State (i) the interest period, (ii) compound interest, (iii) compound amount.

2. James Voss is planning to invest $2000. How much will he have on deposit at the end of 3 years if this sum of money is invested at 12% per annum compounded (a) 3 times a year? (b) 4 times a year? (c) 6 times a year?

3. Amy Allen has succeeded in saving $3000 which she now plans to reinvest. How much will she have on deposit at the end of 4 years if she invests this amount at 8% per annum compounded (a) semi-annually? (b) quarterly?

4. The Andrius Company expects that $5000 will be needed in 5 years to meet the cost of equipment replacement. How much should be initially invested at 12% per annum compounded quarterly to meet this expense?

5. The Lee family wants to have $8000 available in 6 years for the education of their daughter Michelle. How much should be initially invested at 8% per annum compounded quarterly to meet this expense?

6. Jason Marks claims that his money will "not quite double" if deposited for 18 years at 4% per annum compounded semi-annually. Is his claim true?

7. Arnold Bloom purchased a TV set by paying $200 down and agreeing to pay $150 in 6 months. If money is worth 8% per annum compounded quarterly, what was the selling price of the TV set?

FUTURE VALUE OF AN ANNUITY

Suppose that $1000 is invested 6 months from now and in the following three successive 6-month periods at the rate of 6 percent per annum compounded semiannually. The time at which each sum of $1000 is invested coincides with the time at which compound interest is added to the existing amount. At the end of the investment period (2 years from now) these respective amounts will have the following values:

$$1000(1.03)^3 = 1000(1.09273) = \$1092.73$$
$$1000(1.03)^2 = 1000(1.06090) = \$1060.90$$
$$1000(1.03)^1 = 1000(1.03000) = \$1030.00$$
$$1000(1.03)^0 = 1000(1.00000) = \$1000.00$$

The growth exhibited by this investment situation is shown in Figure 3.1.

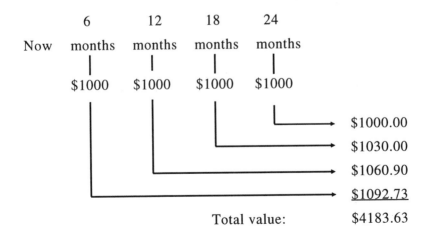

Figure 3.1

Thus the total future value of this investment stream at the end of the investment period is the sum $1092.73 + $1060.90 + $1030.00 + $1000.00 = $4183.63.

This example illustrates a structural feature common to many financial situations:

1. Equal payments R are made at regular intervals of time (in our example, $R = 1000). Such a series of payments is called an **annuity**.

2. Each payment R, called **rent**, is made at the end of the time interval in which it is due. Each such time period is termed a **payment period** (in our example, the payment period is 6 months). Although the term annuity suggests annual payments, the payment period of an annuity may be any length of time.

3. The interest period of the compound interest rate coincides with the payment period of the annuity.

The **term** of an annuity is the time between the beginning of the first payment period and the end of the last payment period. The term of the annuity considered in the preceding example is 2 years. The **amount**, or **future value**, of an annuity is the total amount that would be accumulated at the end of the term of the annuity if each payment is invested at a given rate of compound interest at the time of payment. The future value of the annuity discussed in the preceding example is $F = $4183.63.

More generally, suppose that a rent of R dollars is paid at the end of each payment period for n payment periods, where the compound interest rate is i per period. The R dollars paid at the end of the first payment period will earn interest for $n - 1$ periods, and will amount to $R(1 + i)^{n-1}$ dollars at the end of the term. The R dollars paid at the end of the second payment period will earn interest for $n - 2$ periods, and will amount to $R(1 + i)^{n-2}$ dollars. And so on; the R dollars paid at the end of the $(n - 1)$st (next to last) payment period will earn interest for one period, and will amount to $R(1 + i)$ dollars. The R dollars paid at the end of the nth payment period will earn no interest at all, and will amount to R dollars. In summary, we obtain the time diagram shown in Figure 3.2.

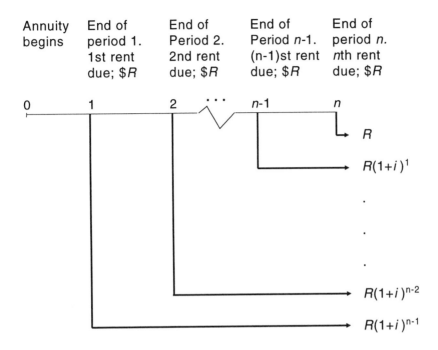

Figure 3.2

Letting F denote the future value of this annuity (the total amount on deposit after n payment periods), we have:

$$F = R + R(1 + i) + R(1 + i)^2 + \ldots + R(1 + i)^{n-2} + R(1 + i)^{n-1}$$

the sum of the first n terms of a geometric progression with first term R and common ratio $(1 + i)$. From the formula for the sum of the first n terms of a geometric progression,

$$S_n = \frac{a(1 - r^n)}{1 - r}$$

with $a = R$ and $r = (1 + i)$ we obtain:

$$F = \frac{R[1 - (1 + i)^n]}{1 - (1 + i)} = \frac{R[1 - (1 + i)^n]}{-i}$$

Multiplying numerator and denominator by -1 and regrouping terms yields:

$$F = R\left[\frac{(1 + i)^n - 1}{i}\right]$$

The quantity $[(1 + i)^n - 1]/i$ is denoted by $s_{\overline{n}|i}$, read **"s angle n with respect to i,"** and expresses the future value of an annuity in which $1 is paid at the end of each payment period for n payment periods, where the compound interest rate is i per period. Thus we have:

$$F = R \cdot s_{\overline{n}|i}$$

Values of $s_{\overline{n}|i}$ for various n and i are given in Table 3.3.

Example 4

To save up enough money for a downpayment on a house, the Roberts family plans to deposit $300 at the end of every 3 months into a fund that pays 8 percent per annum compounded quarterly. How much will be on deposit at the end of 5 years?

This is an annuity with rent R = $300, compound interest of 2 percent per payment period, and 20 payment periods. Thus the amount that will be on deposit at the end of 5 years is $F = 300 \cdot s_{\overline{20}|0.02}$. From Table 3.3, $s_{\overline{20}|0.02} = 24.29737$. Thus:

$$F = 300(24.29737) = \$7289.21$$

Example 5

The Riccardi Company anticipates an expenditure of $6000 to replace a company car in 5 years. How much should be deposited into a fund paying 8 percent per annum compounded semiannually if $6000 is to be available in 5 years?

The interest rate is 4 percent per payment period, the number of payment periods is 10, the amount F is $6000, and the rent R is to be determined. We have $6000 = R \cdot s_{\overline{10}|\,0.04}$, from which we obtain:

$$R = \frac{6000}{s_{\overline{10}|\,0.04}}$$

From Table 3.4, $1/s_{\overline{10}|\,0.04} = 0.08329$. Thus:

$$R = 6000(0.08329) = \$499.74$$

Problems of this sort, for which the amount F and $s_{\overline{n}|\,i}$ are known and the rent R is to be determined, are often referred to as **sinking fund** problems.

Food for Thought

8. Find the amount in a fund at the end of 7 years if the fund pays interest at the rate of 4 percent per annum (a) compounded annually and $50 is deposited at the end of each year; (b) compounded semiannually and $50 is deposited at the end of each 6 months; (c) compounded quarterly and $50 is deposited at the end of each 3 months.

9. At the end of each year Professor Janet Reed deposits 10 percent of her $80,000-a-year teaching salary into a fund paying 5 percent per annum compounded annually. How much money will she have in the fund at the end of 30 years?

10. Oliver Lukas deposits $500 at the end of each year into a fund that pays 4 percent per annum compounded annually. Construct a schedule showing the growth of the fund during the first 5 years.

11. Find the rent required to develop a sinking fund of $2500 in 9 years if deposits are made (a) annually and the interest rate is 5 percent per annum compounded annually; (b) semiannually and the interest rate is 4 percent per annum compounded semiannually; (c) quarterly and the interest rate is 4 percent per annum compounded quarterly.

12. Jack Bryan buys a new car every 2 years. The list price of the new car is $14,000 and the trade-in value of the old car is $7000. If Jack can deposit money into a fund that pays 8 percent per annum compounded quarterly, how much should he deposit each quarter to take care of depreciation on the car?

PRESENT VALUE OF AN ANNUITY

Consider an annuity in which $R is paid at the end of each payment period, where the compound interest rate is i per period. One concern, which we have already explored, is with the future value of the annuity after a certain number of payment periods have elapsed. Another feature, which we now examine, is concerned with the present. What is the annuity worth now? More precisely, what sum initially invested will yield $R at the end of each payment period, where the compound interest rate is i per period? To determine this sum, which, naturally enough, is called the **present value** of the annuity, we determine the present value of each rent R of the annuity and add.

The present value of the first rent of $R, which is due at the end of the first period, is $R(1 + i)^{-1}$. The present value of the second rent is $R(1 + i)^{-2}$, and so on. The present value of the nth rent (which is the last payment) is $R(1 + i)^{-n}$. In summary, we obtain the time diagram shown in Figure 3.3.

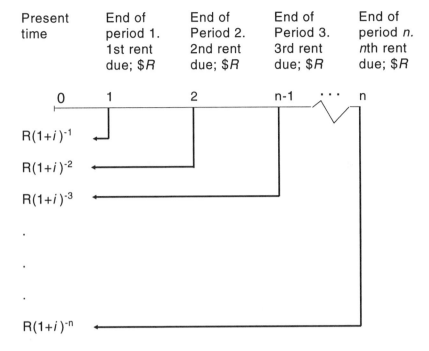

Figure 3.3

Letting P denote the present value of this annuity, we have,

$$P = R(1 + i)^{-1} + R(1 + i)^{-2} + R(1 + i)^{-3} + \ldots + R(1 + i)^{-n},$$

the sum of the first n terms of a geometric progression with first term $R(1 + i)^{-1}$ and common ratio $(1 + i)^{-1}$. From the formula for the sum of the first n terms of a geometric progression,

$$S_n = \frac{a(1 - r^n)}{1 - r}$$

with $a = R(1 + i)^{-1}$ and $r = (1 + i)^{-1}$, we obtain:

$$P = \frac{R(1 + i)^{-1} \, [1 - [(1 + i)^{-1}]^n]}{1 - (1 + i)^{-1}}$$

$$= \frac{R(1 + i)^{-1} \, [1 - (1 + i)^{-n}]}{1 - (1 + i)^{-1}}$$

Multiplying numerator and denominator by $(1 + i)$ neutralizes $(1 + i)^{-1}$ [since $(1 + i)(1 + i)^{-1} = (1 + i)^0 = 1$], and yields:

$$P = \frac{R[1 - (1 + i)^{-n}]}{(1 + i) - 1}$$

$$P = R\left[\frac{1 - (1 + i)^{-n}}{i}\right]$$

The quantity $[1 - (1 + i)^{-n}]/i$ is denoted by $a_{\overline{n}|\,i}$, read **"a angle *n* with respect to *i*,"** and expresses the present value of an annuity in which $1 is paid at the end of each payment period for n payment periods, where the compound interest rate is i per period. Thus we have:

$$P = R \cdot a_{\overline{n}|\,i}$$

Values of $a_{\overline{n}|\,i}$ are given in Table 3.5.

Example 6

An annuity pays $600 every 4 months for 10 years, where the interest rate is 9 percent per annum compounded 3 times a year. Find the present value of this annuity.

The interest rate i is 3 percent per payment period, the number of payment periods is $n = 30$, and the rent is $R = \$600$. From Table 3.5, $a_{\overline{30}|\,0.03} = 19.60044$. Thus we have:

$$P = 600 \cdot a_{\overline{30|} \, 0.03} = 600(19.60044) = \$11,760.26$$

$11,760.26 should be deposited initially into a fund paying interest at the rate of 9 percent per annum compounded 3 times a year if $600 is to be paid every 4 months for 10 years.

Example 7

The Blyden Company has a $10,000 mortgage, which is to be paid off in equal installments to be made quarterly for 5 years. If the interest rate is 8 percent per annum compounded quarterly, what is the amount due in each installment?

The interest rate is 2 percent per payment period, and the number of payment periods is 20. The present value of the annuity is $P = \$10,000$, the amount that is due now but is to be paid off in equal installments. The amount of each installment R is the rent to be determined. From $P = R \cdot a_{\overline{n|} \, i}$, we obtain:

$$R = \frac{P}{a_{\overline{n|} \, i}} = \frac{10.000}{a_{\overline{20|} \, 0.02}}$$

$1/a_{\overline{20|} \, 0.02} = 0.06116$. Thus:

$$R = 10,000(0.06116) = \$611.60$$

Let's take another look at the structure underlying the Blyden Company's problem, but cast it in terms of a different setting.

Example 8

Jennifer Rice purchased a new car for $14,000. She paid $4,000 down and obtained financing for the $10,000 balance (mortgage, if you will) which required the debt to be paid off quarterly in equal installments over five years. The interest is 8 percent per annum compounded quarterly. How much is each installment (rent, if you will)?

Since Jennifer's situation, structurally speaking, is the same as the Blyden Company's, the answer is the same: $611.60.

Problems of this sort for which P and $a_{\overline{n}|i}$ are known and R is to be determined are called **amortization problems.**

Example 9

Apart from taxes and miscellaneous fees, how much did the car actually cost Jennifer? What interest did she incur?

As we saw in Example 8, Jennifer's loan of $10,000 is to be paid in 20 installments of $611.20, which comes to:

$$20 \, (611.60) = \$12,232$$

Thus, the cost of the car comes to $16,232 with $2,232 being the interest paid on the loan.

Example 10

What percent of the total cost of the car is interest?

$$\frac{\text{interest}}{\text{total cost}} \cdot 100 = \frac{2,232}{16,232} \cdot 100 = 13.75\%$$

Example 11

Let us suppose that on reconsideration Jennifer decides not to take out the $10,000 loan, but to put the $611.60 installments that she would have had to pay into a fund paying the same rate of 8 percent per annum compounded quarterly. How long would it take her to accumulate the additional $10,000 to purchase the car?

To answer this question consider a sequence of $611.60 payments from the point of view of an annuity. The basic relationship giving the amount (or future value) of this sequence of payments is:

$$F = R \cdot s_{\overline{n}|\,0.02}$$

In this situation F = $10,000, R = $611.60, i = 0.02 and n, the number of 3-month payment periods, is to be determined. We have:

$$10,000 = (611.60)s_{\overline{n}|\,0.02}$$

$$s_{\overline{n}|\,0.02} = \frac{10,000}{611.60}$$

$$= 16.3506$$

Our next step is to run our finger down the 2% column of Table 3.3 for $s_{\overline{n}|\,i}$ until we locate the smallest value exceeding 16.3506. This value is 17.29342 for which n = 15.

Thus, it will take Jennifer 15 3-month periods or 3 years and 9 months to accumulate the additional $10,000 needed to purchase the car, assuming that the price hasn't changed.

Food for Thought

13. An annuity pays $400 every 3 months for 10 years, where the interest rate is 12 percent per annum compounded quarterly. Find the present value of this annuity.

14. An annuity pays $1000 every 6 months for 8 years, where the interest rate is 8 percent per annum compounded semiannually. Find the present value of this annuity.

15. An annuity pays $100 every month for 3 years, where the interest rate is 12 percent per annum compounded monthly. Find the present value of this annuity.

16. If money is worth 12 percent per annum compounded monthly, which television set is cheaper and by how much?

   a. Set *A* is paid for with a $55 downpayment followed by 12 monthly payments of $10 each.

   b. Set *B* is paid for with a $25 downpayment followed by 15 monthly payments of $10 each.

17. A car costing $850 is paid for with a downpayment followed by 12 monthly payments of $40 each. How much is the downpayment if the interest rate on the unpaid balance is 12 percent per annum compounded monthly?

18. An $8000 mortgage is to be paid off in equal installments to be made at the end of the year for 4 years.

   a. If the interest rate is 5 percent per annum compounded annually, find the amount paid in each installment.

   b. Make a schedule showing the part of each installment that goes for interest and principal.

19. To pay off a mortgage of $6000 in 5 years, what should the rent be if the installments are payable (a) annually at 4 percent per annum compounded annually? (b) semiannually at 4 percent per annum compounded semiannually? (c) quarterly at 4 percent per annum compounded quarterly?

20. Amy Ho will need college expenses of $1500 on each of her 18th, 19th, 20th, and 21st birthdays. What single sum of money should be deposited on Amy's 14th birthday in a fund paying 4 percent per annum compounded annually to provide the necessary college funds?

21. William Baxter, who is 55, plans to deposit equal sums of money in a retirement fund every 6 months until he is 65 years old, when the fund has $10,000 in it. He arranges for the money to be paid to him (or his heirs) in equal amounts at the end of every 6 months for 10 years. Money is worth 6 percent per annum compounded semiannually.

a. Before retirement, how much must be deposited into the fund at the end of each 6-month period?

b. Upon retirement, how much does he receive at the end of each 6-month period?

22. At the end of each year for 10 years, Susan Rey deposits $200 in a bank that pays interest at the rate of 5 percent per annum compounded annually. The money continues to draw interest until the 15th year, at which time the sum on deposit is withdrawn. How much is withdrawn?

23. Karen Duran borrows $2000 at 12 percent per annum compounded monthly and agrees to pay off the debt in equal monthly payments for 2 years. After she makes her 15th payment, she decides to pay off the balance of the debt. How much should she pay?

24. Joseph Carter purchases a compact disk player and agrees to pay for it in 24 equal monthly payments of $50 each. Money is worth 12 percent per annum compounded monthly. After he makes his 20th payment, he decides to pay off the balance of the debt. How much should he then pay?

25. Consider an annuity for which a rent of R is paid at the beginning of each payment period for n payment periods, where the interest rate is i per period. Show that the future value, or amount, of this annuity is $F = (1 + i)R \cdot s_{\overline{n}|i}$, and that its present value is $P = (1 + i)R \cdot a_{\overline{n}|i}$.

26. Julio Monteiro purchased a new car for $22,000. He paid $5000 down and obtained financing for the $17,000 balance. The agreement called for the debt to be paid off semi-annually in equal installments over eight years. The interest rate is 10 percent per annum compounded semi-annually.

a. How much is each installment?

b. How much did the car actually cost Julio, apart from taxes and miscellaneous fees?

c. What interest did he incur?

d.   What percent of the total cost of the car is interest?

e.   Suppose that on reconsideration Julio decides not to take out the $17,000 loan, but to put the installments that he would have had to pay into a fund paying the same rate of 10 percent per annum compounded semi-annually. How long would it take him to accumulate the additional $17,000 to purchase the car, assuming that its price hasn't changed?

Table 3.1 Compound Interest: $(1 + i)^n$

n	1%	2%	3%	4%	5%
1	1.01000	1.02000	1.03000	1.04000	1.05000
2	1.02010	1.04040	1.06090	1.08160	1.10250
3	1.03030	1.06121	1.09273	1.12486	1.15762
4	1.04060	1.08243	1.12551	1.16986	1.21551
5	1.05101	1.10408	1.15927	1.21665	1.27628
6	1.06152	1.12616	1.19405	1.26532	1.34010
7	1.07214	1.14869	1.22987	1.31593	1.40710
8	1.08286	1.17166	1.26677	1.36857	1.47746
9	1.09369	1.19509	1.30477	1.42331	1.55133
10	1.10462	1.21899	1.34392	1.48024	1.62889
11	1.11567	1.24337	1.38423	1.53945	1.71034
12	1.12683	1.26824	1.42576	1.60103	1.79586
13	1.13809	1.29361	1.46853	1.66507	1.88565
14	1.14947	1.31948	1.51259	1.73168	1.97993
15	1.16097	1.34587	1.55797	1.80094	2.07893
16	1.17258	1.37279	1.60471	1.87298	2.18287
17	1.18430	1.40024	1.65285	1.94790	2.29202
18	1.19615	1.42825	1.70243	2.02582	2.40662
19	1.20811	1.45681	1.75351	2.10685	2.52695
20	1.22019	1.48595	1.80611	2.19112	2.65330
21	1.23239	1.51567	1.86029	2.27877	2.78596
22	1.24472	1.54598	1.91610	2.36992	2.92526
23	1.25716	1.57690	1.97359	2.46472	3.07152
24	1.26973	1.60844	2.03279	2.56330	3.22510
25	1.28243	1.64061	2.09378	2.66584	3.38635
26	1.29526	1.67342	2.15659	2.77247	3.55567
27	1.30821	1.70689	2.22129	2.88337	3.73346
28	1.32129	1.74102	2.28793	2.99870	3.92013
29	1.33450	1.77584	2.35657	3.11865	4.11614
30	1.34785	1.81136	2.42726	3.24340	4.32194
31	1.36133	1.84759	2.50008	3.37313	4.53804
32	1.37494	1.88454	2.57508	3.50806	4.76494
33	1.38869	1.92223	2.65234	3.64838	5.00319
34	1.40258	1.96068	2.73191	3.79432	5.25335
35	1.41660	1.99989	2.81386	3.94609	5.51602
36	1.43077	2.03989	2.89828	4.10393	5.79182
37	1.44508	2.08069	2.98523	4.26809	6.08141
38	1.45953	2.12230	3.07478	4.43881	6.38548
39	1.47412	2.16474	3.16703	4.61637	6.70475
40	1.48886	2.20804	3.26204	4.80102	7.03999

Table 3.2 Present Value: $(1 + i)^{-n}$

n	1%	2%	3%	4%	5%
1	.99010	.98039	.97087	.96154	.95238
2	.98030	.96117	.94260	.92456	.90703
3	.97059	.94232	.91514	.88900	.86384
4	.96098	.92385	.88849	.85480	.82270
5	.95147	.90573	.86261	.82193	.78353
6	.94205	.88797	.83748	.79031	.74622
7	.93272	.87056	.83109	.75992	.71068
8	.92348	.85349	.78941	.73069	.67684
9	.91434	.83676	.76642	.70259	.64461
10	.90529	.82035	.74409	.67556	.61391
11	.89632	.80426	.72242	.64958	.58468
12	.88745	.78849	.70138	.62460	.55684
13	.87866	.77303	.68095	.60057	.53032
14	.86996	.75788	.66112	.57748	.50507
15	.86135	.74301	.64186	.55526	.48102
16	.85282	.72845	.62317	.53391	.45811
17	.84438	.71416	.60502	.51337	.43630
18	.83602	.70016	.58739	.49363	.41552
19	.82774	.68643	.57029	.47464	.39573
20	.81954	.67297	.55368	.45639	.37689
21	.81143	.65978	.53755	.43883	.35894
22	.80340	.64684	.52189	.42196	.34185
23	.79544	.63416	.50669	.40573	.32557
24	.78757	.62172	.49193	.39012	.31007
25	.77977	.60953	.47761	.37512	.29530
26	.77205	.59758	.46369	.36069	.28124
27	.76440	.58586	.45019	.34682	.26785
28	.75684	.57437	.43708	.33348	.25509
29	.74934	.56311	.42435	.32065	.24295
30	.74192	.55207	.41199	.30832	.23138
31	.73458	.54125	.39999	.29646	.22036
32	.72730	.53063	.38834	.28506	.20987
33	.72010	.52023	.37703	.27409	.19987
34	.71297	.51003	.36604	.26355	.19035
35	.70591	.50003	.35538	.25342	.18129
36	.69892	.49022	.34503	.24367	.17266
37	.69200	.48061	.33498	.23430	.16444
38	.68515	.47119	.32523	.22529	.15661
39	.67837	.46195	.31575	.21662	.14915
40	.67165	.45289	.30656	.20829	.14205

Table 3.3 Future Value of an Annuity: $s_{\overline{n}|i}$

n	1%	2%	3%	4%	5%
1	1.00000	1.00000	1.00000	1.00000	1.00000
2	2.01000	2.02000	2.03000	2.04000	2.05000
3	3.03010	3.06040	3.09090	3.12160	3.15250
4	4.06040	4.12161	4.18363	4.24646	4.31012
5	5.10101	5.20404	5.30914	5.41632	5.52563
6	6.15202	6.30812	6.46841	6.63298	6.80191
7	7.21354	7.43428	7.66246	7.89829	8.14201
8	8.28567	8.58297	8.89234	9.21423	9.54911
9	9.36853	9.75463	10.15911	10.58280	11.02656
10	10.46221	10.94972	11.46388	12.00611	12.57789
11	11.56683	12.16872	12.80780	13.48635	14.20679
12	12.68250	13.41209	14.19203	15.02581	15.91713
13	13.80933	14.68033	15.61779	16.62684	17.71298
14	14.94742	15.97394	17.08632	18.29191	19.59863
15	16.09690	17.29342	18.59891	20.02359	21.57856
16	17.25786	18.63929	20.15688	21.82453	23.65749
17	18.43044	20.01207	21.76159	23.69751	25.84037
18	19.61475	21.41231	23.41444	25.64541	28.13238
19	20.81090	22.84056	25.11687	27.67123	30.53900
20	22.01900	24.29737	26.87037	29.77808	33.06595
21	23.23919	25.78332	28.67649	31.96920	35.71925
22	24.47159	27.29898	30.53678	34.24797	38.50521
23	25.71630	28.84496	32.45288	36.61789	41.43048
24	26.97346	30.42186	34.42647	39.08260	44.50200
25	28.24320	32.03030	36.45926	41.64591	47.72710
26	29.52563	33.67091	38.55304	44.31174	51.11345
27	30.82089	35.34432	40.70963	47.08421	54.66913
28	32.12910	37.05121	42.93092	49.96758	58.40258
29	33.45039	38.79223	45.21885	52.96629	62.32271
30	34.78489	40.56808	47.57542	56.08494	66.43885
31	36.13274	42.37944	50.00268	59.32834	70.76079
32	37.49407	44.22703	52.50276	62.70147	75.29883
33	38.86901	46.11157	55.07784	66.20953	80.06377
34	40.25770	48.03380	57.73018	69.85791	85.06696
35	41.66028	49.99448	60.46208	73.65222	90.32031
36	43.07688	51.99437	63.27594	77.59831	95.83632
37	44.50765	54.03425	66.17422	81.70225	101.62814
38	45.95272	56.11494	69.15945	85.97034	107.70955
39	47.41225	58.23724	72.23423	90.40915	114.09502
40	48.88637	60.40198	75.40126	95.02552	120.79977

Table 3.4 $\dfrac{1}{s_{\overline{n}|i}}$

n	1%	2%	3%	4%	5%
1	1.00000	1.00000	1.00000	1.00000	1.00000
2	.49751	.49505	.49261	.49020	.48780
3	.33002	.32675	.32353	.32035	.31721
4	.24628	.24262	.23903	.23549	.23201
5	.19604	.19216	.18835	.18463	.18097
6	.16255	.15853	.15460	.15076	.14702
7	.13863	.13451	.13051	.12661	.12282
8	.12069	.11651	.11246	.10853	.10472
9	.10674	.10252	.09843	.09449	.09069
10	.09558	.09133	.08723	.08329	.07950
11	.08645	.08218	.07808	.07415	.07039
12	.07885	.07456	.07046	.06655	.06283
13	.07241	.06812	.06403	.06014	.05646
14	.06690	.06260	.05853	.05467	.05102
15	.06212	.05783	.05377	.04994	.04634
16	.05794	.05365	.04961	.04582	.04227
17	.05426	.04997	.04595	.04220	.03870
18	.05098	.04670	.04271	.03899	.03555
19	.04805	.04378	.03981	.03614	.03275
20	.04542	.04116	.03722	.03358	.03024
21	.04303	.03878	.03487	.03128	.02800
22	.04086	.03663	.03275	.02920	.02597
23	.03889	.03467	.03081	.02731	.02414
24	.03707	.03287	.02905	.02559	.02247
25	.03541	.03122	.02743	.02401	.02095
26	.03387	.02970	.02594	.02257	.01956
27	.03245	.02829	.02456	.02124	.01829
28	.03112	.02699	.02329	.02001	.01712
29	.02990	.02578	.02211	.01888	.01605
30	.02875	.02465	.02102	.01783	.01505
31	.02768	.02360	.02000	.01686	.01413
32	.02667	.02261	.01905	.01595	.01328
33	.02573	.02169	.01816	.01510	.01249
34	.02484	.02082	.01732	.01431	.01176
35	.02400	.02000	.01654	.01358	.01107
36	.02321	.01923	.01580	.01289	.01043
37	.02247	.01851	.01511	.01224	.00984
38	.02176	.01782	.01446	.01163	.00928
39	.02109	.01717	.01384	.01106	.00876
40	.02046	.01656	.01326	.01052	.00828

Table 3.5 Present Value of an Annuity $a_{\overline{n}|i}$

n	1%	2%	3%	4%	5%
1	0.99010	0.98039	0.97087	0.96154	0.95238
2	1.97040	1.94156	1.91347	1.88609	1.85941
3	2.94099	2.88388	2.82861	2.77509	2.72325
4	3.90197	3.80773	3.71710	3.62990	3.54595
5	4.85343	4.71346	4.57971	4.45182	4.32948
6	5.79548	5.60143	5.41719	5.24214	5.07569
7	6.72819	6.47199	6.23028	6.00205	5.78637
8	7.65168	7.32548	7.01969	6.73274	6.46321
9	8.56602	8.16224	7.78611	7.43533	7.10782
10	9.47130	8.98259	8.53020	8.11090	7.72173
11	10.36763	9.78685	9.25262	8.76048	8.30641
12	11.25508	10.57534	9.95400	9.38507	8.86325
13	12.13374	11.34837	10.63496	9.98565	9.39357
14	13.00370	12.10625	11.29607	10.56312	9.89864
15	13.86505	12.84926	11.93794	11.11839	10.37966
16	14.71787	13.57771	12.56110	11.65230	10.83777
17	15.56225	14.29187	13.16612	12.16567	11.27407
18	16.39827	14.99203	13.75351	12.65930	11.68959
19	17.22601	15.67846	14.32380	13.13394	12.08532
20	18.04555	16.35143	14.87747	13.59033	12.46221
21	18.85698	17.01121	15.41502	14.02916	12.82115
22	19.66038	17.65805	15.93692	14.45112	13.16300
23	20.45582	18.29220	16.44361	14.85684	13.48857
24	21.24339	18.91393	16.93554	15.24696	13.79864
25	22.02316	19.52346	17.41315	15.62208	14.09394
26	22.79520	20.12104	17.87684	15.98277	14.37519
27	23.55961	20.70690	18.32703	16.32959	14.64303
28	24.31644	21.28127	18.76411	16.66306	14.89813
29	25.06579	21.84438	19.18845	16.98371	15.14107
30	25.80771	22.39646	19.60044	17.29203	15.37245
31	26.54229	22.93770	20.00043	17.58849	15.59281
32	27.26959	23.46833	20.38877	17.87355	15.80268
33	27.98969	23.98856	20.76579	18.14765	16.00255
34	28.70267	24.49859	21.13184	18.41120	16.19290
35	29.40858	24.99862	21.48722	18.66461	16.37419
36	30.10751	25.48884	21.83225	18.90828	16.54685
37	30.79951	25.96945	22.16724	19.14258	16.71129
38	31.48466	26.44064	22.49246	19.36786	16.86789
39	32.16303	26.90259	22.80822	19.58448	17.01704
40	32.83469	27.35548	23.11477	19.79277	17.15909

4 ▶ ARE WE ON THE RIGHT DATA TRAIL?

4.1 FOOD FOR THOUGHT

1. Abel Fisher, a financial analyst, was hired to analyze the operations of the accounting department of Arley College and make recommendations on how to improve its efficiency. The income of the department is the tuition income of the students being serviced minus costs, primarily salary costs. Mr. Fisher collected data on the class size of each instructor and each instructor's rank and salary. He found that a number of the full professors at the

top of the salary scale were teaching classes with a small number of students. To improve the income of the department he recommended that teachers at the top of the salary scale be assigned classes which can be expected to have a large number of students.

a. Is the data collected appropriate to the problem under study? Explain.

b. Do you agree that implementation of Abel Fisher's recommendation will improve the accounting department's efficiency from an income-cost point of view? Explain.

2. The Baldwin Insurance Company hired marketing analyst Arnold W. Williamson to develop a strategy to make its car insurance policies more attractive. Mr. Williamson suggested that to reward and encourage safe driving the company offer a 5% discount to policy holders who had been with the company for five years and had not been in an accident. An additional 1% discount would be given for each additional year that had been accident free up to a maximum of 15%. Do you agree that the number of years of safe driving is the number to focus on as a measure of safe driving?

3. Table 4.1 gives the number of accidents involving deaths, serious passenger injuries or substantial damage to plane for nine airlines in the most recent five year period.

Table 4.1

Airline	A	B	C	D	E	F	G	H	I
Accidents	14	12	12	10	9	5	3	2	2

Based on this data Janet Reed, a spokeswoman for airline H, claimed that no airline had a better safety record than H; "fly H for safety" became the airline's motto. Is this data a suitable measure of airline safety?

4. "Don't cancel the course in Modern Lithuanian Literature," said the Chairman of the East European Languages Department to the Dean of Academic Affairs of Ecap University. "The course reg-

istration increased by 25% over last week and with such rapid growth I'm optimistic that we'll have a strong registration," he further pointed out. Is the percentage growth in the course's enrollment the figure that should be focused on in making a decision on whether or not to run the course?

5. The Association for the Advancement of Business Research (AABR) gives its accreditation to colleges and universities that have a strong business research dimension. Ecap University applied for AABR accreditation, citing as evidence of its business research dimension the fact that 600 papers were published by its business faculty over the last three years for an average of two papers per faculty member per year. Is this the kind of data that an evaluation team should focus on as an indication of a strong business research dimension?

6. "I don't know what you're complaining about," said the baseball owners' representative to the players' counterpart. "The players make, on average, $1.2 million." Is the average salary the best indicator of player salaries?

7. Which data should we use to determine the highest taxed state? For discussion see: T. Redburn, "Report Says New York Taxes Are the 2nd Highest in the U.S.," *The New York Times*, Sept. 21, 1994; B1.

8. President G. Marx of Huxley College charged his Dean of Administrative Affairs, Humphrey Appleby, with the task of setting up a criterion for running or canceling course sections which would take into account student needs, be sensitive to maintaining academic quality, address the cost dimension, and be simple to use.

Dean Appleby started with the assumption that each section, with perhaps a few exceptions, should pay its own way. He set up a course section run-cancel criterion in terms of the difference between the tuition revenue generated for the section based on student enrollment and the salary cost of the instructor for the section. The Appleby criterion, HA-1, is to run the section if revenue minus cost exceeds $5000. Otherwise, it is to be canceled, unless a compelling student need for the course could be established.

a.   What data would have to be collected to implement HA-1?

b.   Each faculty member at Huxley College is required to teach 7 courses (or sections) during the academic year. Student tuition is $1,000 per course. Professor Edith White, a recognized authority on American poetry, who earns $84,000 for the academic year, is scheduled to teach English 205, Modern American Poetry. One week before the course was to begin it was noted that 16 students had registered for ENG 205. If a decision had to be made on whether to run or cancel ENG 205 at this point, what would the decision be under HA-1?

c.   What is the minimum student enrollment that would be needed to run ENG 205?

d.   It came to pass that the head of the English department had to make some program changes and Professor White was replaced by Professor Brian Roberts as the instructor for ENG

205. Professor Roberts, who is at the beginning of his teaching career, is paid $42,000 for the academic year. Would ENG 205 run now?

e. Generally speaking, how vulnerable is the implementation of HA-1 to changes in teaching assignments?

f. How well does HA-1 and the data implicit in its implementation satisfy President Marx's conditions?

9. Dean Appleby was asked to develop another criterion for running or canceling course sections which takes into account student needs, is sensitive to maintaining academic quality, addresses the cost dimension, and is simple to use. The dean came up with HA-2, which operates as follows: For a given department, English, for example, determine the average salary cost of the faculty in the English department. Run English section A, let us say, if the tuition revenue based on section A's enrollment exceeds the average salary cost of the English department for section A by $5000. Otherwise, cancel section A unless compelling student or academic needs can be established. The same system would apply to courses in other departments.

a. What data would be needed for the implementation of HA-2?

b. Let us suppose that the English department of 10 faculty has an average salary of $70,000 for the academic year, during which each faculty member is required to teach 7 courses. Student tuition is $1000 per course. If 16 students were registered for ENG 205, Modern American Poetry, would it run under HA-2?

c. How does HA-2 compare with HA-1?

d. Are there any disadvantages to HA-2?

e. How well does HA-2 and the data implicit in its implementation satisfy President Marx's conditions.

10. HA-3, developed by Dean Appleby, operates as follows: For a given department, determine the average class size for all sections being run by the department for the session in question,

excluding such one-on-one activities as independent study and thesis supervision. If this average class size is at least 25, then run all sections with an enrollment of at least 10. If the enrollment of a section is less than 10, then go to HA-2 to make a determination for the section. If the average class size is less than 25, then go to HA-2 for all sections. Address the same questions for HA-3 as those posed for HA-2 in 9. (In 9c replace HA-1 by HA-2.)

11. You have been hired as a consultant to help Dean Appleby develop a criterion for running and canceling course sections which addresses the conditions stated by President Marx (see 8).

    a. Describe the run/cancel criterion that you would recommend.

    b. What data would have to be obtained to implement your criterion?

    c. What are the merits of your criterion compared to HA-1, HA-2, and HA-3?

    d. Are there any disadvantages to your criterion?

12. At this point, with 2130 consecutive games played, Lou Gehrig holds baseball's endurance record. Barring injuries and illness, Cal Ripken stands poised to break this record in early September 1995 and succeed Gehrig as the "Iron Horse." Number of consecutive games played is a striking number, but is it the best measure of endurance? Is the number of consecutive innings in the number of consecutive games played a better measure? (Ripken, for example, played over 900 games without missing an inning. Gehrig played consecutive innings in every game in one season.) How about position played? (Ripken's position of shortstop leaves a player more prone to injury than Gehrig's position of first base, some would argue.)

13. America's total expenditure on health in 1991 as a percentage of gross domestic product was 13.3%, the world's highest, according to the United Nations' 1994 Human Development Report. This report also reports America's per capita expenditure on health as $2932 for 1991, the world's highest. Some would argue that these figures mean that America has the best health care in the world. Are they on the right data trail?

14. The figures make clear that America's health care system is a failure, some critics argue. Over 14% of the population lack health insurance. Americans' life expectancy (at birth, 1992) ranks behind 17 other nations. As to health care delivery, the average hospital stay of Americans admitted to hospitals was 6.5 days compared with 12.9 days in industrial countries for which data were available (1989). Are they on the right data trail?

15. Concerned about charges of subtle patterns of bias at its executive levels, The United Federation of Worlds set up a Commission to investigate. During its hearings Lork from Mork pointed out that while Morkians are 30% of the Federation's work force at its lower levels, they make up only 1% of its executive staff. "Good faith recruitment efforts have been made," observed Lork, "but the statistics show subtle patterns of discrimination against Morkians." Tallia from Talos I disagreed. "The statistics show discrepancies," countered Tallia. "The subtle patterns of discrimination are your interpretation of the discrepancies. Look at the Universe Games held every hundred years. For the last two thousand years 75% of the participants chosen by the trials have been Morkians. Does this mean that the trials were biased in favor of the Morkians?" "Certainly not," answered Lork, "They earned the right to be there in the trials that were held."

    a. What do the statistics tell us?

    b. How would you interpret them?

    c. Is Mork on the right data trail in trying to determine if bias exists?

16. In "Statistics Reveal Bulk of New Jobs Pay Over Average" (*The New York Times*, Oct. 17, 1994; p. A1) Sylvia Nasar reports that most of the 5.5 million jobs generated by the economy in the two and a half years since the beginning of 1992 pay more than the average wage. The average hourly pay has increased by 2.5% (after inflation) from $13.44 in 1990 to $15.43 by mid-1994. Average total hourly compensation increased by about 3.5% (after inflation) from $15.09 to $17.63 during the same period.

Would it be justified to conclude that the economy has turned around and that workers as a group are sharing in the economic gains?

17. Should you allow, assuming that you have the authority, your closet friends and relatives to be Democrats and marry other Democrats? Consider the following data reported by R. Morin (*The Washington Post National Weekly Edition*, Sept. 26–Oct. 2, 1994; p. 37) based on survey data obtained by the University of Chicago's National Opinion Research Center between 1972 and 1993. The results are due to political scientist Lee Sigelman who compared answers given by Democrats and Republicans to questions about their lives.

    a. *Democrats are poor.* The average income of Democrats was 25% below that of Republicans.

    b. *Democrats have unpleasant habits.* 37% of them smoke, as compared to 30% of Republicans; 23% of them have seen an X-rated movie within the past year, as compared to 17% of Republicans.

c. *Democrats are unhappy.* 31% of them describe themselves as very happy, as compared to 39% of Republicans.

What does the future hold for Democrats? From the data, some would suggest a dismal future: poverty, disreputable behavior, unhappiness, with inbreeding generating more and more of the same. What is your reaction to this scenario? Explain in appropriate detail.

18. Intelligence is loosely defined as the capacity to reason abstractly, to solve problems, to cope with one's environment, to organize large amounts of information into useful systems. It is generally agreed that both heredity and environment play a role in intelligence. Developing tests to obtain quantitative measures of intelligence is another matter, because with any such proposed test there is an underlying question of how well the test expresses what it is intended to measure. Some psychologists would operationally define intelligence in terms of the particular test employed. Whatever intelligence tests measure, it is correct to note that there is a correlation between I.Q. scores and such factors as school performance and success on the job.

    Consider a classification of people into three categories, I's, D's and R's, and let us suppose that the I's, as a group, tend to score several points higher on an I.Q. test than the D's and that the D's, as a group, tend to score several points higher than the R's. This raises several questions.

    a. Is it "legitimate" to conclude from this that the I's are superior in intelligence to the D's who in turn are superior in intelligence to the R's?

    b. Does the data establish that there are hereditary differences in intelligence for the I's, D's, and R's?

    c. Is it "legitimate" to conclude that the R's are condemned to be at the "bottom" of society?

    d. Is I.Q. really everything?

These are not new issues, but they have received renewed attention from the publication of [5], [6] and [12] and a number of critiques and commentaries, some of which are noted below.

REFERENCES

1. W. Allman, "Why IQ Isn't Destiny," *U.S. News & World Report*, Oct. 24, 1994; pp. 73–80.

2. P. Brimelow, "For Whom the Bell Tolls," *Forbes*, Oct. 24, 1994; pp. 153–158.

3. M. Browne, "What is Intelligence, and Who Has It?" *The New York Times Book Review*, Oct. 16, 1994; p. 3.

4. S. Fraser, ed. *The Be,, Curve Wars: Race, Intelligence and the Future of America* (New York: Basic Books, 1995).

5. R. Herrnstein, C. Murray. *The Bell Curve: Intelligence and Class Structure in American Life* (New York: The Free Press, 1994).

6. S. Itzkoff. *The Decline of Intelligence in America: A Strategy for National Renewal* (Westport, Conn.: Praeger, 1994).

7. R. Jacoby, N. Glauberman, ed., *The Bell Curve Debate* (New York: Times Books, 1995).

8. J. Leo, "Return to the IQ Wars," *U.S. News & World Report*, Oct. 24, 1994; 24.

9. P. Passell, Review of [5], *The New York Times*, Oct. 27, 1994; p. C19.

10. P. Passell, "Bell Curve Critics Say Early I.Q. Isn't Destiny," *The New York Times*, Nov. 9, 1994; p. A25.

11. W. Raspberry, "Is I.Q. Really Everything" *The Washington Post National Weekly Review*, Oct. 17–23, 1994; p. 29.

12. J.P. Rushton. *Race, Evolution, and Behavior: A Life History Perspective* (New Brunswick, N.J.: Transaction Publishers, 1994).

13. T. Sowell, "Ethnicity and IQ," [4; pp. 70–79].

COMING UP WITH THE DATA

5.1 POLLS AND QUESTIONNAIRES

Food for Thought

1. State and discuss in appropriate detail the problems that arise from newspaper polls and their modern equivalents.

2. Does increasing the sample size by itself insure greater reliability of the poll to be taken? Explain.

3. The City Council of Bell City wants to obtain a sense of the public's view of the effectiveness of the City's Medical Emergency Response Team (MERT). A random sample of 200 residents was chosen from the City's home owners listing and sent a questionnaire. One hundred twenty five responses were received; 95 gave MERT a favorable rating and 30 gave MERT an unfavorable rating. The City Council concluded that MERT has a 76% favorable rating by the public and, with much satisfaction, announced this result to the news media. Do you agree or disagree with their conclusion. Explain.

4. Mayor Keith Joos of Masters-on-the-Mississippi is planning to run for re-election. On a recent talk show he invited the public to call a 900 number and express a favorable or unfavorable rating of his administration. Five hundred calls were received; 350

callers gave his honor an unfavorable rating and 150 gave him a favorable rating. With a 70% unfavorable rating, he strongly considered not running for re-election. Is his pessimism over the results of this call-in warranted? Explain.

5. "I don't understand what went wrong," lamented an editor of *The Literary Digest*. "We were so on-target with our 1932 poll and so off-target with our 1936 poll, for which we used the same methods." What would you tell him?

6. What factors contributed to the failure of the Crossley, Gallup, and Roper polls to correctly predict the outcome of the 1948 presidential election? Explain.

7. Many hotels have a practice of leaving post card questionnaires in guests' rooms inviting comments and ratings. Discuss the pros and cons of this method for obtaining guests' reactions.

8. How might exit surveys in presidential elections be seriously flawed? (See [21].)

9. The Center for Critical Thinking on Domestic and World Affairs is preparing a questionnaire to obtain a sample of public opinion on domestic and world affairs. The following is a sample of some of the questions being posed.

    a. Would you vote for a presidential candidate who was willing to take more out of your pocket by raising taxes?

    b. Do you want the nation's defense capability reduced by budget cuts in an age of rampant terrorism?

    c. Do you believe that very high priority should be given to reducing the crushing budget deficit that has been imposed on our country?

    d. Do you believe that we should continue to squander money on foreign aid while there are so many urgent domestic needs that require attention?

Two response options were to be allowed: Yes, No. The Center is engaged in pre-testing its questionnaire. If you were approached, what opinion would you give them about the four questions noted and the response options?

10. The following questions are from a survey circulated by the Democratic National Committee in September 1995. Do you have suggestions on how the wording of any of these questions or the response options could be modified to obtain a more accurate reflection of a respondent's beliefs without leading the respondent?

a. Do you favor or oppose Republican plans to make huge cuts in entitlement spending, including $270 billion in reductions in Medicare which will force the average Social Security recipient to spend almost half of his or her COLA just to cover resulting higher out-of-pocket Medicare costs?

   ( ) favor   ( ) Oppose   ( ) Undecided

b. Meanwhile the Republicans have proposed a "new budget plan" which would provide large tax cuts for the rich, large increases in defense spending and cuts in social programs and entitlements. Republicans claim that their plan will reduce the deficit. What is your opinion?

   ( ) The Republican plan would raise the national debt.

   ( ) The Republican plan would lower the national debt.

   ( ) The Republican plan would make no difference.

c. Do you favor or oppose Republican proposals to dismantle Head Start and programs that provide health care and nutrition to young mothers in poverty?

   ( ) Favor   ( ) Oppose   ( ) Undecided

d. Republican leaders are advocating a welfare reform plan that would end welfare benefits to single parents who cannot find work. Do you support this approach?

   ( ) Favor   ( ) Oppose   ( ) Undecided

e. President Clinton has proposed a balanced budget plan that doesn't place the burden on the backs of our nation's senior citizens. His plan protects Social Security, ensures the future of Medicare as a part of health care reform, and calls for real cuts in other areas of government—while opposing tax cuts for the wealthiest Americans. Do you favor this plan?

( ) Favor   ( ) Oppose   ( ) Undecided

f. President Clinton is supporting welfare reform that will require single parents to accept work and provide child care and job training during a transitional period. If no work is available, participants will be expected to perform community service work for their benefits. Do you favor or oppose this plan?

( ) Favor   ( ) Oppose   ( ) Undecided

g. Do you favor or oppose legislation to limit special interest campaign contributions and place caps on the amount of money that can be spent in campaigns for federal offices?

( ) Favor   ( ) Oppose   ( ) Undecided

h. Do you favor or oppose the Freedom of Choice Act which, if passed, would secure the right to choose for all women in our nation?

( ) Favor   ( ) Oppose   ( ) Undecided

11. The following question appeared on a survey of American views on the Holocaust: "Does it seem possible or does it seem impossible to you that the Nazi extermination of the Jews never happened?" Comment on the clarity of this question. (See [10].)

12. Is what people say in public opinion polls necessarily what they mean? How can we close the gap between what they say and mean? (See [27].)

13. Andy Arunas of *The Huxley College Press* has been asked to conduct a survey of Huxley's students (estimated at 6000) on

education matters concerned with the quality of the education they are receiving at Huxley.

    a. What questions would you suggest that he include in the questionnaire? Explain.

    b. What response options would you suggest that he allow? Explain.

14. Andy Arunas was also asked by his editor to conduct a survey of Huxley's students to determine the facilities they would prefer in the new student union building that is being planned. Same questions (a) and (b) posed for 13.

15. The Census Bureau is planning a number of changes in its methods for collecting data. What are the changes and what prompted them? (See [9].)

16. How many battered women are there? Are the techniques used for obtaining the data reliable? (See [3] & [12].)

17. In April 1993 the Battelle Memorial Institute of Human Affairs Research published a study on male sexual behavior which stated that 1% of the men surveyed considered themselves exclusively homosexual. This figure differs considerably from the 10% value published in the Kinsey report in 1948.

    a. Discuss the difficulties in obtaining an accurate grip on this number.

    b. Discuss the political dimensions of the different figures.

For discussion see [1], [2], [5], [11], [13], [14], [15], [16], [23], and [25].

18. *The Social Organization of Sexuality* by E. Laumann, J. Gagnon, R. Michael, and S. Michaels (The University of Chicago Press, 1994) and its less technically demanding version *Sex in America* by R. Michael, J. Gagnon, E. Laumann and G. Kolata (Little Brown & Co., 1994) have been hailed as the most definitive study of American sexual behavior to date. Ninety minute interviews were conducted with 3432 people between ages 18 and 59 ran-

domly chosen from a cross section of American households, which constituted a high response rate of nearly 80%. The great strength of the study, it has been observed, is that the random sampling techniques employed make it possible to project its findings onto the population at large with a high degree of confidence. Still, are there reasons to be cautious about extending its findings onto the population at large? For discussion see [13], [14], [15], [16], [18], and [26].

19. The history of pre-election polling shows that polling results often swing wildly before the decisive verdict of election day. In other cases they have been known to settle consistently on one candidate only to be upstaged on election day by the victory of the favored candidate's main competitor. A number of questions thus arise. Are wild swings in polls due to differences in sampling procedures, revisions of opinion in the electorate, or random sampling error? Are poll differences largely due to differences in the way in which pollsters ask questions? What does it say about the reliability of polls if they consistently indicate that a candidate is leading, but he loses the election? See [11], [17], [19], [22], and [27].

20. What should we look for to help us distinguish polls that have been properly carried out from their bogus cousins?

21. In the pioneering days of polling George Gallup expressed the view that polling provided a reliable means of going directly to the people to determine their views. "The modern poll can, and to a certain extent does, function as a creative arm of government." [8; p. 151] Others found the idea simplistic at best and preposterous at worst. Not only are there sharp limitations on the questions that can be posed but it is absurd to base government on answers to individual questions without giving proper consideration to the overall whole (see, for example, [24]). And then there is the view that modern polling has been used more and more to manipulate public opinion than express it (see [19]). Discuss these points of view in an essay. What is your opinion?

22. The following questions were among those asked in a *New York Times*/CBS News Poll conducted Oct. 22–24, 1995 on How the Public Views Budget Choices (see [4]). House Speaker Newt

Gingrich and other Republican leaders attacked the poll as containing deliberately rigged questions calculated to prompt unfavorable public reaction to Republican proposals (see [6]).

If you had to choose would you prefer:	% Response
1 a. Balancing the Federal budget, or	27
b. Preventing Medicare from being significantly cut.	67
2 a. Balancing the Federal budget, or	60
b. Cutting taxes.	35
3 a. Balancing the Federal budget, or	21
b. Preventing Social Security from being significantly cut.	71

Concerning 1, Republican officials argued that the word "cut" was inflammatory and did not accurately reflect the Republican plan which, they contended, slow projected growth of Medicare. Concerning 2, Republican officials argued that their plan would accomplish both objectives. Concerning 3, it was argued that Republican plans do not touch Social Security.

Joseph Lelyveld, executive editor of *The Times,* replied that the questions had been carefully worded and that long-established polling techniques had been employed. Neither the wording of the poll nor the results differ significantly from what other news and polling organizations found in recent months, he also noted.

As far as can be decided from the information provided by [4], [6], and other sources that you can obtain, which view is closer to the mark—rigged questions that are totally phony or carefully worded question where long-established polling techniques were followed? Discuss.

REFERENCES

1. F. Barringer, "Sex Survey of American Men Finds 1% Are Gay," *The New York Times*, April 15, 1993; A1.

2. F. Barringer, "Polling on Sexual Issues Has Its Drawbacks," *The New York Times*, April 25, 1993; D23.

3. A. Brott, "The Facts Take a Beating," *The Washington Post National Weekly Edition*," Aug. 8–14, 1994; pp. 24–25.

4. A. Clymer, "Americans Reject Big Medicare Cuts, A New Poll Finds, *The New York Times,* Oct. 26, 1995; A1.

5. D. Dunlap, "Gay Survey Raises a New Question," *The New York Times*, Oct. 18, 1994; B8.

6. I. Fisher, "Gingrich Attacks Times–CBS Poll, Claiming Bias Against GOP," *The New York Times*, Oct. 27, 1995; D21.

7. G. Gallup, S.F. Rae, *The Pulse of Democracy* (New York: Greenwood Press, 1968).

8. G. Gallup, "Opinion Polling in a Democracy," *Statistics: A Guide to the Unknown*, ed. by J.M. Tanur, et al. (Holden Day, 1972), 146–152.

9. S. Holmes, "Census Officials Plan Big Changes in Gathering Data," *The New York Times*, May 16, 1994; A1.

10. J. Kifner, "Pollster Finds Error on Holocaust Doubts," *The New York Times*, May 5, 1994.

11. L. Kusmin, "When Polls Swing, Don't Blame Fickle Voters," Letter, *The New York Times*, Sept. 26, 1992.

12. J. Leo, "Is It a War Against Women?" *U.S. News & World Report*, July 11, 1994; 22.

13. R. C. Lewontin, "Sex, Lies and Social Science," *The New York Review of Books*, April 20, 1995; 24–29.

14. R. C. Lewontin et. al., "Sex, Lies, and Social Science: An Exchange," *The New York Review of Books*, May 25, 1995; 43–44.

15. R. C. Lewontin et. al., "Sex, Lies, and Social Science: Letters," *The New York Review of Books*, June 8, 1995; 68–69.

16. R. C. Lewontin et. al., "Sex, Lies, and Social Science: Another Exchange," *The New York Review of Books*, Aug. 10, 1995; 55–56.

17. M. Lipset, "Polls Don't Lie. People Do," *The New York Times*, Sept. 10, 1992.

18. M. G. Lord, "What that Survey Didn't Say," *The New York Times*, Oct. 25, 1994; A22.

19. D. Moore, *The Super Pollsters: How They Measure and Manipulate Public Opinion in America* (New York: Four Walls Eight Windows, 1992).

20. D. Moore, "The Sure Thing that Got Away," *The New York Times*, Oct. 25, 1992.

21. R. Morin, "When the Data Tell Shockingly Different Stories," *The Washington Post National Weekly Edition*, Aug. 8–13, 1994; 37.

22. F. Newport, "Look at Polls as a Fever Chart of the Electorate," *The New York Times*, Nov. 6, 1992.

23. P. Painton, "The Shrinking Ten Percent," *Time,* April 23, 1993; 28–29.

24. L. Rogers, *The Pollsters: Public Opinion, Politics, and Democratic Leadership* (Alfred A. Knopf, 1949).

25. J. Schmalz, "Survey Stirs Debate on Number of Gay Men in U.S." *The New York Times*, April 16, 1993; A19.

26. J. K. Wilson, "U.S. Sex Survey Had Flawed Methodology," *The New York Times*, Nov. 1, 1994; A26.

27. D. Yankelovich, "What Polls Say—and What They Mean," *The New York Times*, Sept. 17, 1994; 23.

5.2 SURVEY OF SAMPLING METHODS

In class, on the job, and in statistics books, among others, we hear or read comments of the following sort: Take a sample of size 30 from the population; survey public opinion on this or that issue. Easily said, but far from easily done. In this section we briefly look at some methods of sampling and problems that arise in carrying them out.

Sampling methods fall into two general classes, **probability sampling**, which includes random sampling (or simple random sampling, as it is sometimes called), systematic selection, stratified sampling, and cluster sampling, and **nonprobability sampling**, which includes quota sampling and judgment sampling.

In the following we assume that the underlying population is finite, that is, that the number of elements or units in the population can be described by a positive integer; the integer might be small, such as 2, or fairly large, such as 2 trillion.

RANDOM SAMPLING

A sample of a certain size, 10 units, let us say, is to be chosen from the underlying population. When we say that the sample is to be chosen at **random** we have in mind the idea that there is to be no bias, deliberate or inadvertent, which favors certain samples of size 10 being chosen over others. The sampling procedure is to be an equal opportunity procedure. Whether the randomly chosen sample is to be of size 10 or 2 or 1500 is immaterial; the sample is to be chosen in an unbiased manner; no favoritism.

Andy Arunas, freshman reporter for *The Huxley College Press*, wants to interview a sample of Huxley's faculty concerning their views on students, administration, and the issues of cheating, grading, tenure, teaching, and research. Andy is considering choosing a random sample of 10 of Huxley's 300 faculty. To do this he is thinking of proceeding in the following way.

1. Obtain a list of the 300 faculty and number them 1 through 300;

2. Obtain 300 ping-pong balls from the local sport equipment store and number them 1 through 300;

3. Put the balls in a bag, shake the bag, turn it upside down several times, and then (without peeking) draw 10 balls from the bag, one at a time. The faculty corresponding to the 10 numbers drawn would be faculty Andy would seek to interview if he decides to choose his interviewees by random sampling.

Does this device ensure that random sampling will be carried out? It is easy to declare that random sampling is what we desire or assume to be the case, but achieving it in practice is another matter.

SYSTEMATIC SELECTION WITH A RANDOM START

The idea of systematic sampling is to choose every twentieth name on a list, or every hundredth number in a telephone directory, or every fifth house in a residential neighborhood. To add an element of randomness to this procedure one may choose the starting point at random.

If Andy were considering systematic selection with a random start as a mechanism for choosing a sample of 10 of the 300 faculty, he would begin by determining the **sampling interval** k by taking the ratio of the population size to the sample size and rounding to the nearest integer. Andy's sampling interval is k = 300/10 = 30. The next step is to choose an integer at random from 1, 2, . . ., 30. Suppose 11 were obtained; then Andy would seek to interview the faculty numbered 11, 11 + 30 = 41, 11 + 2(30) = 71, . . ., 11 + 9(30) = 281 should he decide on this approach.

An appealing feature of this sampling procedure is that it is easy to carry out and spreads the sample through the population. It may yield unrepresentative results, however. If every thirtieth faculty number on the list beginning with the eleventh comes from Huxley's Business School, for example, the sample obtained would consist entirely of business faculty.

If the underlying population is listed at random, then systematic selection with a random start is equivalent to random sampling.

STRATIFIED RANDOM SAMPLING

Sometimes a population can, with respect to a defining characteristic, be "meaningfully" divided into relatively homogeneous subpopulations or *strata* such that each member of the population belongs to one and only one *stratum*. Figure 5.1 shows a population Q which has been partitioned into five *strata* denoted by $S_1, S_2, \ldots, S_5$.

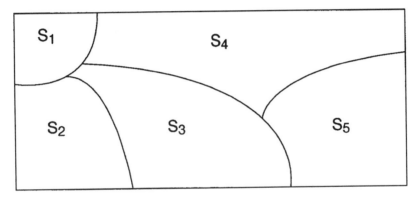

Figure 5.1

Stratified random sampling consists of "meaningfully" partitioning the population into *strata* and then taking a random sample from each *stratum*. To employ stratified random sampling a listing must be available for each *stratum*. The overall sample size n may or may not be allocated to the *strata* in proportion to the *strata* sizes. If n is allocated in this way, the sampling is called **proportional sampling**.

Andy Arunas, for example, might consider partitioning the Huxley College Faculty according to school: Arts and Sciences, Business, Education, Mathematics and Statistics, and Equine Studies. With an envisioned total sample size of 10, he might follow up by choosing 2 faculty at random from each school, but this is not his only option.

If all the schools have the same number of faculty, then the subsample allocation of 2 per school constitutes proportional sampling.

Stratified random sampling is effective in reducing sampling error when the population is diverse, appropriate *strata* can be defined which are relatively homogeneous, and it is desirable to have each *stratum* represented in the overall sample. An interesting readily accessible application of stratified sampling to a problem in accounting involving the Chesapeake and Ohio Freight Study is discussed by John Neter in [1].

CLUSTER RANDOM SAMPLING

In cluster random sampling the population is partitioned into relatively small subunits called **clusters** (rather than *strata*). A random selection of clusters is chosen and a subsample (possibly 100 percent) of each chosen cluster is taken.

Huxley College's Faculty are organized into departments (accounting, art, economics, etc.) which may be thought of as clusters. To select the faculty to be interviewed by cluster sampling using the departments as clusters, Andy would proceed by choosing a sample of departments at random and then interviewing a sample of the faculty from each of the departments from the sample chosen.

Estimates based on random sampling are generally more reliable than those based on cluster random sampling, when both can be carried out, but cluster sampling is generally more cost effective to implement and is sometimes a viable option when random sampling is not feasible. Suppose, for example, that an organization wants to study family income levels in the Boston area and has decided to interview 1500 families. Random sampling may not be feasible because a complete list of the families (the target population) may not be available. Moreover, the cost of interviewing families scattered over a large area may be prohibitively high. An alternative approach would be to divide the region into city blocks, the clusters in this case, and interview all or a sample of families in a number of randomly selected blocks.

In some applied situations it may be necessary to use a combination of the aforediscussed sampling procedures. If an education organization, for example, wanted to study education problems faced by elementary school teachers in the United States, its statisticians might proceed by stratifying by states, partitioning the states into counties, and using cluster sampling to obtain a sample of counties. The last stage might involve using random sampling to emerge with schools whose personnel or sample of personnel would be interviewed.

NONPROBABILITY SAMPLING

Nonprobability sampling includes any sampling method which does not include probability tools in its design. Two main methods of nonprobability sampling are judgment sampling and quota sampling.

JUDGMENT SAMPLING

As its name suggests, a **judgment sample** is one selected on the basis of someone's judgment. If Andy decides to interview those faculty who he had as instructors, then this would clearly be judgment sampling.

Judgment sampling is useful in carrying out a small scale survey and in pilot studies which precede major surveys. A judgment sample might involve the choice of a few "typical" cities to test market a new product. It might involve choosing a small sample on which to try out a questionnaire to detect unforeseen difficulties prior to launching a major survey.

As is clear from its nature, if the judgment exercised in choosing the sample is good, the results will be insightful; if poor judgment is exercised, the results might have disastrous consequences.

QUOTA SAMPLING

In a **quota sample** the interviewer is instructed to interview an assigned number, or quota, of individuals in groups defined by specified characteristics. Andy, for example, might be told by his editor to interview 3 faculty from the history department, 2 from the

English department, 2 from the economics department, and 3 from the marketing department. Which faculty should be interviewed is left to Andy's discretion, as long as the above quota is satisfied. Quota sampling is a form of judgment sampling. It is vulnerable to the interviewer's biases in the selection of interviewees. The interviewer may focus on those who are readily available and seek to avoid those with an "unsavory" look, factors which are not controlled by a quota. Random sampling, on the other hand, is merciless in that the interviewer has no choice; if an individual has been pinpointed through a random sample, it is the interviewer's obligation to locate that party, whether the trail is difficult or not, whether the party is unsavory or not.

Quota sampling has sometimes been used in marketing surveys and public opinion polling. It is cheaper per sample unit than random sampling and, when carefully controlled, can yield good results.

Inherent in probability sampling is the means for objectively estimating the precision of the sample results, which is its major advantage over nonprobability sampling.

REFERENCES

1. J. Neter, "How Accountants Save Money by Sampling," *Statistics: A Guide to the Unknown*, ed. J.M. Tanur et. al. (San Francisco: Holden-Day, Inc., 1972), 203–211.

Food for Thought

1. a. What is the difference between probability and nonprobability sampling methods?

   b. Are probability methods always preferable to nonprobability methods? Explain.

2. What is the difference between stratified and cluster sampling?

3. What is the difference between cluster and quota sampling?

4. Describe two situations in which stratified sampling would be advantageous.

5. Andy Arunas has been asked to interview a sample of 10 members of Huxley College's administration about the challenges and problems of administration. Huxley's administration consists of five major divisions: (1) Administrative Affairs, which numbers 15 and includes the president, provost, 8 who are concerned with internal affairs of the college (finance, purchasing, and human resources, for example) and 5 who are concerned with external affairs of the college (fund raising, for example); (2) Academic Affairs, which numbers 45 and includes 5 deans of the earlier mentioned schools, 35 academic department heads, and 5 directors of special programs (honors, international studies, etc.); (3) Athletics, which numbers 3; (4) Student Services, which numbers 10 (bursar, registrar, financial aid director, etc.); and (5) Academic Support Services, which numbers 7 (directors of library, computer services, etc.)

a. Describe in reasonable detail how Andy might implement the probability and nonprobability sampling methods discussed in this section to carry out his task.

b. Discuss the advantages and disadvantages of these methods.

6. The faculty, administration, and other workers at Huxley College may be viewed as consisting of four non-overlapping groups: Faculty (300), Administration (80), Supervisory staff (20), Non-supervisory staff (150). A sample of those working at Huxley is to be drawn and their average salary used as an estimate of the average salary of all who work at the college. The question is, how should the sample be drawn? Discuss the merits and drawbacks of random sampling, stratified sampling, and quota sampling for this problem.

7. To carry out a marketing survey on consumer preferences for kitchen appliances Elias Marketing Research Associates placed two interviewers on the busiest street in town to interview passersby. Does the sample of opinions obtained qualify as a random sample? Explain.

8. *The Huxley College Press*, which wants to obtain a sense of faculty opinion about the College's budget allocations for the forthcoming year, sent a reporter to the School of Business to interview a randomly chosen sample of faculty in the accounting, finance, marketing, management, and business law departments about this matter. Will doing this give the *Press* what it wants? Explain.

9. To obtain a sense of the business community's opinion of what ails American business, *Business Tomorrow* interviewed a randomly chosen sample of CEO's of Fortune 500 companies. Will doing this give *Business Tomorrow* what it desires? Explain.

10. The Harold Institute for Urban Studies is seeking to obtain a sense of the opinion of Baxter City's residents about the problems of the poor. To obtain this they interviewed a random sample of Baxter City's residents whose names appeared in a list of those who voted in last year's municipal elections. Will doing this give the Harold Institute what it seeks? Explain.

11. Tickets were sold at Ecap University's graduation celebration to help raise funds for the University's new library. The tickets sold were placed in a bowl as soon as they were sold. At the end of the graduation festivities the University's Library Director, Harriet Warren, reached into the bowl and chose a ticket at random. The ticket holder was awarded a newly published edition of Charles Dickens' collected works. Kevin Reynolds, who was among the first to purchase a ticket protested that the drawing procedure was biased and demanded that ticket purchasers be given a refund or that the drawing be held again. "Your claim is not justified Mr. Reynolds," replied the Dean of Student Affairs. "Ms. Warren was blindfolded and the ticket was chosen at random." Who is right? Explain.

VALIDITY VERSUS TRUTH

6.1 DEDUCTIVE REASONING, VALIDITY, AND TRUTH

To simply illustrate the nature of a valid conclusion consider the two statements

1. All x's are y's.

2. All y's are z's.

and the statement

3. All x's are z's.

Suppose we take statements (1) and (2) as a starting point for the purpose of seeing what conclusions follow as a logical consequence. In general, a collection of statements set down for such a purpose is called an **hypothesis**. Each statement in an hypothesis is referred to, variously, as an **assumption**, **premise**, **postulate**, or **axiom**, depending on context.

The **proof** or **argument** which establishes that a purported conclusion does indeed follow as an inescapable consequence of the postulates is said to be a **valid proof** or **valid argument** and the conclusion of a valid proof is said to be **valid with respect to the postulates**, or **hypothesis** made up of the postulates. Valid conclusions of an hypothesis are called **theorems**.

These are basic definitions which require flushing out if we are to get a secure hold on them. To begin this process we return to our mini-system consisting of (1), (2), and (3).

Postulate P1: All x's are y's.

Postulate P2: All y's are z's.

Conclusion C1: All x's are z's.

These statements assert relationships between various classes of objects identified as x's, y's, and z's. P1 forces the class of x's within the class of y's, with P2 forcing the class of y's within the still more inclusive class of z's. It follows as an inescapable consequence that the x's, like it or not, are forced within the z's, which is the content of C1. Conclusion C1 is valid with respect to postulates P1 and P2, or put another way, C1 is valid with respect to the hypothesis consisting of P1 and P2; it is a theorem in this system and we may upgrade it from C1 to T1, for theorem.

A convenient way to picture these relationships is by means of diagrams of the following sort. Represent each class by points inside a closed curve. Thus, P1 is represented by placing the class of x's within the class of y's as shown in Figure 6.1(a). P2 is represented by placing the class of y's within the class of z's as shown in Figure 6.1(b). The diagrams help us to get a better grip on the assumed relationships between the x's, y's, and z's.

The x's, y's, and z's are "abstract" entities in the fullest sense of the term and it is important to note that T1 is a valid consequence of P1 and P2, irrespective of the nature of the x's, y's, and z's. The validity of T1 from P1 and P2 is a structural condition, irrespective of the content assigned to the x's, y's and z's. It's somewhat like having a person in the flesh—this is, the real person—who we may dress up in many ways—business suit, beach wear, evening attire, what have you. The person's appearance changes, sometimes radi-

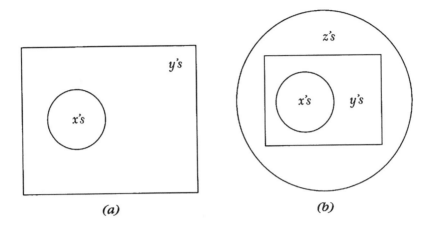

Figure 6.1

cally, but it is fundamentally the same person. So it is with valid conclusions. We may color-in x, y, and z in many ways, but doing so changes the appearance of P1, P2, and T1 and not the validity of T1 on the basis of P1 and P2.

Suppose, for example, that we add color to x, y, and z as follows. Let:

x = apple, y = people, z = animal

This gives us the hypothesis:

P1a: All apples are people.

P2a: All people are animals.

and the valid conclusion:

T1a: All apples are animals.

But T1a is wrong, you say. Well, yes and no; we must be very careful to distinguish the sense in which T1a is wrong from the sense in which it is correct. T1a is correct in the sense of being valid with respect to P1a and P2a; it is wrong, that is, false, as a statement about the relationship between apples and animals. Adding the coloring x = apple, y = people, z = animal to the scene does not change anything about the structure of the argument which determines validity, but it does introduce a truth/falsity dimension into the scene which, of course, complicates it.

A system made up of postulates and theorems is called, as one would expect, a **postulate system**. The process of obtaining valid conclusions from the postulates of the system is called **deduction** or **deductive reasoning**. A postulate system I obtained from an "abstract" postulate system APS by assigning representations to its abstract terms (x, y, z or equivalents) is called an **interpretation** of the APS. Validity relationships are maintained in passing from an "abstract" postulate system APS (such as P1, P2, and T1) to any interpretation of APS (such as I_1 consisting of P1a, P2a, T1a).

If we let

$$x = dog, y = mammal, z = animal$$

we obtain interpretation I_2 with postulates

P1b: All dogs are mammals.

P2b: All mammals are animals.

and theorem

T1b: All dogs are animals.

If we let

x = apple, y = cat, z = fruit

we obtain interpretation I_3 with postulates

P1c: All apples are cats.

P2c: All cats are fruit.

and theorem

T1c: All apples are fruit.

The interpretations I_1, I_2, and I_3 illustrate the following relationships between the validity of a statement and its truth.

1. If T is a theorem in an interpretation I of an APS and the hypothesis of I is true, that is, its postulates are true, then T is true. Valid conclusions deduced from true postulates are true.

This is illustrated by I_2. It makes sense since in obtaining a valid conclusion T of an hypothesis H we do not go beyond H. If H is true and we do not go beyond it to obtain T, then it is not surprising that T is also true.

2. If the theorems of an interpretation I of an APS are true, then we cannot conclude that the hypothesis H of I is true. H might be true; some of the postulates of H might be true and others false; it might be false in its entirety.

I_3 illustrates an interpretation with a true theorem arising from false postulates.

3. If a theorem T of an interpretation I of an APS is false; then some of the postulates of I must be false.

This third property follows from the fact that if the postulates of I were true, then we could not obtain from them a false theorem T. I_1 illustrates an interpretation with a false theorem (T1a: All apples are animals) arising from a system with a false postulate (P1a: All apples are people).

In summary then, we have:

1. If the postulates are true, then the theorems must be true.

2. If the theorems are true, then the postulates may or may not be true.

3. If a theorem is false, then some of the postulates must be false.

INVALID ARGUMENTS

Consider the following structure.

Hypothesis.   P1: All x's are y's.

P2: All z's are y's.

Conclusion.   C1: All x's are z's.

A diagrammatic representation of P1 and P2 is shown in Figure 6.2. The argument is not valid, or **invalid,**

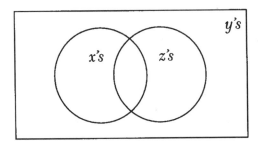

Figure 6.2

because C1 is not forced by the hypothesis. We are forced by P1 to place the class of x's within the class of y's and by P2 to place the class of z's within the class of y's, but this does not force us to place the x's within the z's. It may be that the x's are contained within the z's, but "may be" isn't strong enough for the argument to be valid.

The mini-systems considered here enable us to initiate a discussion of validity versus truth. For an adequate perspective we must consider a richer framework, which is taken up in the next section.

Food for Thought

Determine the validity of the arguments for the proposed conclusions in 1–9. Explain the basis for your answer.

1. Hypothesis.   P1: All mammals are frogs.

   P2: Joe Warren is a mammal.

   Conclusion.   C1: Joe Warren is a frog.

2. Hypothesis.   P1: All oranges are blueberries.

   P2: All blueberries are fruit.

   Conclusion.   C1: All oranges are fruit.

3. Hypothesis.   P1: Some college students are geniuses.

                    P2: All freshmen are geniuses.

   Conclusions.   C1: All freshmen are college students.

                    C2: No freshmen are college students.

                    C3: Some freshmen are college students.

                    C4: Some college students are freshmen.

4. Hypothesis.   P1: No college students are geniuses.

                    P2: All freshmen are college students.

   Conclusion.   C1: No freshmen are geniuses.

5. Hypothesis.   P1: No math professors are bores.

                    P2: Some math professors are human.

   Conclusion.   C1: Some humans are not bores.

6. Hypothesis.   P1: All babies are beautiful.

                    P2: Amy is beautiful.

   Conclusion.   C1: Amy is a baby.

7. Hypothesis.   P1: All dogs are mammals.

                    P2: Some mammals are frogs.

                    P3: Some mammals are white.

   Conclusions.   C1: Some dogs are frogs.

                    C2: Some frogs are white.

                    C3: Some dogs are white.

8. Hypothesis.   P1: Some x's are y's.

                     P2: All x's are z's.

   Conclusions.   C1: Some z's are y's.

                     C2: Some z's are not y's.

9. Hypothesis.   P1: All x's are y's.

                     P2: Some x's are z's.

                     P3: Some w's are x's.

   Conclusions.   C1: Some y's are z's.

                     C2: Some y's are w's.

                     C3: Some z's are w's.

                     C4: Some z's are not w's.

                     C5: Some z's are not x's.

                     C6: Some z's are not y's.

10. What is it about the "wisdom" stated in the cartoon that prompted the reply, "What have you guys been drinking?"

6.2 DUBIOUS DEDUCTIONS

"What did you learn in geometry?" asked Jeff Arnold of his daughter Jenny. "I don't remember," replied Jenny. "Well, then, what did you do in geometry?" countered Mr. Arnold. "Most of the time we were doing proofs," answered Jenny.

Deductive proofs, which define the backbone of mathematics, are first encountered by most of us in a course in geometry. The development of geometry, called Euclidean geometry after the Greek geometer Euclid of Alexandria (c. 300 B.C.), begins with a statement of some basic postulates. Its theorems are then logically deduced from the postulates and, as theorems begin to accumulate, from a mixture of postulates and previously proved theorems. In developing a proof each statement in the sequence of statements defining the proof is justified by a statement within the system itself (postulate, theorem, construction, definition). Diagrams play a fundamental role in guiding us to the conclusion we seek to reach.

But sometimes the diagrams take on a life of their own and our orderly sequence of logical inferences are broken by employing visual evidence which, strictly speaking, go beyond the legitimate means of justification that we have available to us in the system. The following two examples illustrate the difficulties that arise.

Purported theorem. There exists a triangle with two right angles.

Proof. Consider two circles that intersect at two points which we shall term A and B (see Figure 6.3). Let $\overline{AC}$, $\overline{AD}$ denote their respective diameters from A. Let

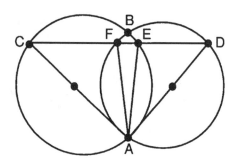

Figure 6.3

CD meet the respective circles in E, F. Thus angle AEC is a right angle, since it is inscribed in semicircle AEC. Similarly, angle AFD is a right angle. Thus triangle AEF has two right angles.

The argument seems airtight, but something is wrong since a triangle, with angle sum 180 degrees, cannot have two right angles. The diagram is compelling, but a very carefully drawn diagram might suggest that CD passes through B, so that AEF is not a triangle at all but a line segment. While a very carefully drawn diagram might suggest this, there is nothing in the postulates or theorems of Euclidean geometry which allows us to argue that CD passes through B.

The following is a well known theorem of Euclidean geometry, but let us look carefully at the proof that is usually presented.

Theorem. The base angles of an isosceles triangle are equal.

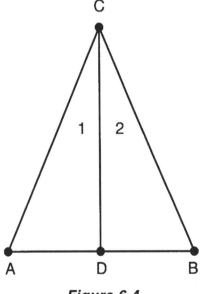

Figure 6.4

Given: Triangle ABC with $\overline{AC} = \overline{BC}$.

To prove: Angle A = Angle B.

Proof.

Statement	Justification
1. Draw the bisector of angle C.	1. Every angle has a bisector.
2. Extend it to meet AB at D.	2. A line may be extended.
3. In triangles ACD and BCD, AC=BC.	3. Hypothesis.
4. Angle 1 = angle 2.	4. Definition of angle bisector.
5. $\overline{CD} = \overline{CD}$	5. Identity.
6. Triangle ACD is congruent to triangle BCD.	6. Side-Angle-Side.
7. Angle A equals angle B.	7. Corresponding parts of congruent triangles are equal.

The proof and diagram are convincing, but is it, by itself, a valid argument? Strictly speaking, we would have to say no. A difficulty arises in Step 2. The justification, a line may be extended, does not say that a line may be extended to meet another line, AB in this case; the lines might be parallel, and we need something in the system itself that would rule this out.

Let us suppose that we can get around this difficulty and conclude that the bisector of angle C does intersect line AB at D. But where is D? For the proof to hold up we need D to fall between A and B, but there is nothing in the system that says it must. Figure 6.4 is so convincing on this point that our first reaction might be to say that it's "obvious" that D is between A and B; where else could it be? "Obvious" is not the same as deductive proof and postulates and theorems on "betweeness" are needed to support such conclusions.

What is the place of diagrams, then? Should we abandon them? The answer to this last question is a resounding NO, NO, NO. Diagrams are invaluable for suggesting ideas and providing us with a sense of what we want to do and what we are obtaining. They are essential allies; but their use should not be equated to formal deductive proof, which must come out of the underlying system itself.

Food for Thought

Pinpoint the gaps in the following proofs.

1. Playfair's form of the parallel postulate: If given a line L and point P not on L, then there is one and only one line which passes through P and is parallel to L. (see Figure 6.5)

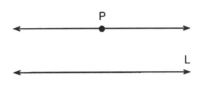

Figure 6.5

Proof.

1. Drop a perpendicular from point P to line L (see Figure 6.6). To this perpendicular erect a perpendicular PE from the point P. This second perpendicular is parallel to line L by the theorem that two perpendiculars to the same line are parallel. Since it is possible to drop only one perpendicular from a given point to a given line, and it is possible to erect only one perpendicular to a line from a point lying on it, the parallel line PE is unique.

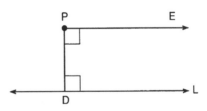

Figure 6.6

2. If a line L intersects two parallel lines M and N, the sum of the interior angles lying on the same side of L is 180 degrees (see Figure 6.7).

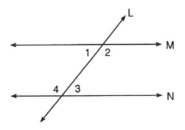

Figure 6.7

Proof.

Three cases are possible:

1. The sum of the interior angles on the same side of L exceeds 180 degrees.

2. The sum of the interior angles on the same side of L is less than 180 degrees.

3. The sum of the interior angles on the same side of L equals 180 degrees.

In the first case we have:

$$\text{angle } 1 + \text{angle } 4 > 180°, \quad \text{angle } 2 + \text{angle } 3 > 180°$$

Thus:

$$\text{The angle sum of 1 through 4 exceeds } 360°$$

But the sum of the four interior angles is equal to two straight angles, 360 degrees. This contradiction shows that Case 1 is untenable. The same reasoning shows that Case 2 is untenable. Thus we are left with Case 3, which is what is to be proved.

Theorem. The diagonals of a parallelogram bisect each other.

Given: Parallelogram ABCD with diagonals $\overline{AC}$ and $\overline{BD}$.

To Prove: $\overline{AE} = \overline{EC}, \quad \overline{BE} = \overline{ED}$

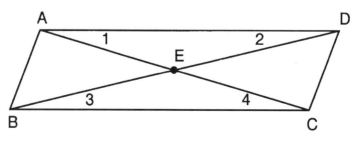

Figure 6.8

Proof.

Statement	Justification
1. AD is parallel to BC; AB is parallel to DC.	1. By hypothesis ABCD is a parallelogram.
2. Angle 1 = Angle 4. Angle 2 = Angle 3.	2. Alternate interior angles of parallel lines are equal.
3. $\overline{AD} = \overline{BC}$.	3. Opposite sides of a parallelogram are equal.
4. Triangles AED and CEB are congruent.	4. Angle-Side-Angle.
5. $\overline{AE} = \overline{EC}$, $\overline{BE} = \overline{ED}$.	5. Corresponding parts of congruent triangles are equal.

4. Return to Exercise 1 for a "proof" by the Greek geometer Proclus (410–485).

From Exercise 1 we have that there exists a line M passing through P parallel to L. (see Figure 6.9).

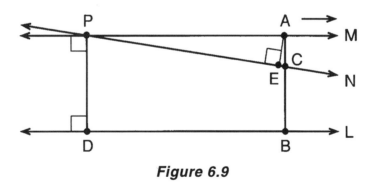

Figure 6.9

The problem is to show that M is unique. Suppose there is another line N through P parallel to L. Then N makes an acute angle with PD that lies on one side or the other of PD. Suppose it's the right side of PD. The part of N to the right of P is thus contained in the region bounded by L, M, and PD.

Let A denote any point of M to the right of P; let AB denote the perpendicular to L at B, and let C denote the point at which AB intersects N. Then $\overline{AB} > \overline{AC}$. Let A recede on M; then $\overline{AC}$ increases without bound since $\overline{AC} \geq \overline{AE}$, the perpendicular from A to line N. Thus AB, which is at least as large as $\overline{AC}$, increases without bound.

But the distance between two parallel lines must be bounded. Thus we have a contradiction to the result that $\overline{AC}$ increases without bound, which means that our supposition that there is another line N through P parallel to L is untenable.

6.3 MORE ON PROOF AND POSTULATE SYSTEMS

To obtain a more complete perspective on proof and postulate systems in mathematics we examine some mini-systems that are more substantial than those considered in Section 6.1.

First, we turn our attention to the system I_a founded on the following postulates.

P1a: Every line is a collection points which contains at least two points.

P2a: For any two points there is at least one line containing them.

P3a: For any line there is a point not contained by it.

P4a: There is at least one line.

We have not defined the basic terms point and line and at first sight it might seem unnecessary to do so. "We all know what point and line indicate," you might say. Yes, most of us think of point as indicating position in space and line as being a path produced by a straight edge of indefinite length, but, as we saw in the preceding section, these associations have the power to cloud our minds when it comes to constructing valid proofs.

To get around this difficulty let us use uncharged terms—zog and glob, for example—keeping in mind that one interpretation that we may give to these terms is point and line. We may also use this envisioned interpretation to draw pictures to help guide our thinking, being careful to keep them at a distance, so-to-speak, so that they do not dominate our thinking to the extent that they close our minds to other possible interpretations.

The abstract postulate system that emerges, call it **Glob Theory**, is based on the following postulates.

P1: Every glob is a collection of zogs which contains at least two zogs.

P2: For any two zogs there is at least one glob containing them.

P3: For any glob there is a zog not contained by it.

P4: There is at least one glob.

What are globs and zogs? They are undefined. Every postulate system begins with undefined terms, undefined in the sense that no unique characterization in terms of more basic entities is given. There is no way around this. If we were to define glob and zog in terms of mumbo and jumbo, let us say, then mumbo and jumbo would be our basic undefined terms. If we sought to define mumbo and jumbo in terms of more basic elements, then they would be our undefined terms, and on it goes. Postulates P1 through P4 do not uniquely define zog and glob, but state some relationships between them.

If we interpret zog and glob as point and line in the usual sense, we emerge with interpretation I_a consisting of postulates P1a through P4a. On the other hand, suppose we interpret zog and glob as follows:

zogs: The points given by the coordinates (1,1), (5,1), (3,5).

globs: The circular arcs joining these points of the circle determined by them (see Figure 6.10).

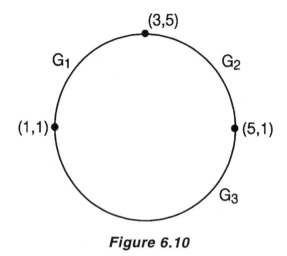

Figure 6.10

There are three globs in this interpretation: G_1, G_2, and G_3.

The postulates in this interpretation of Glob Theory, call it I_b, take the following form:

P1b: Every circular arc is a collection of points which contains at least two points (that is, at least two of (1,1), (5,1), (3,5)).

P2b: For any two points there is at least one circular arc containing them.

P3b: For any circular arc there is a point not contained by it.

P4b: There is at least one circular arc.

For both I_a and I_b it is clear that the conditions stated in the postulates of its parent Glob Theory are satisfied. Such interpretations of an abstract postulate system (APS) are of particular interest; they are called **models** of the parent APS. There are two types of models—concrete and ideal—which we should distinguish between. A model of an abstract postulate system is said to be **concrete** if the interpretation assigned to its undefined terms are objects and relations adopted from the real world; the model is said to be **ideal** if the interpretation assigned to its undefined terms are objects and relations adopted from some other postulate system. I_a and I_b are ideal models of Glob Theory.

A fundamental property of models is that theorems established for the parent abstract postulate system describe relations which hold for any model as well. Since validity is determined by structure and not by how undefined terms are interpreted, theorems proved in the abstract hold for models of an APS as well.

We now turn to proving some theorems to illustrate proof in this sort of setting.

The models I_a and I_b of Glob Theory both contain at least three points, which suggests the possibility that Glob Theory itself contains at least three zogs. Let us undertake to develop a formal proof of this suggested result.

Possibility. Glob Theory contains at least three zogs.

Idea: Get a glob into play since a glob contains at least two zogs. As a visual aid we might refer "informally" to something like Figure 6.11(a). This gives us two zogs, p and q, and we are well under way.

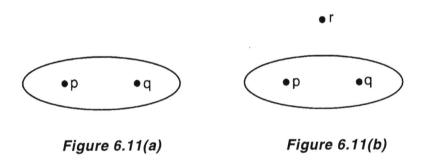

Figure 6.11(a)   *Figure 6.11(b)*

We can nail down our third zog r by appealing to P3. As a visual aid we have Figure 6.11(b). These figures are just that, helpful visual aids; they do not comprise the formal proof.

At this point, with the ideas in hand, we are ready to write down the formal proof.

Theorem 1. Glob Theory contains at least three zogs.

Proof.

Statement	Justification
1. Let L denote a glob.	1. P4
2. L contains at least two zogs, p and q.	2. P1
3. There is a zog r not in L.	3. P3
4. p, q, and r are distinct zogs.	4. Summary of preceding results.

The observation that the zogs are distinct, with justification, is most important. For convenience we have introduced letters to denote the zogs, but the fact that different letters are being used does not by itself guarantee that the zogs are different.

What about the globs? How many can we count on in general? The interpretations I_a and I_b of Glob Theory suggest that there are at least three. Let us see if we can develop a package of ideas to prove this result. A possibility of this sort which is suggested by consideration of special cases is called a **conjecture**.

Conjecture. Glob Theory contains at least three globs.

Idea: we can generate an initial glob, call it L_1, by appealing to P4 (see Figure 6.12(a)).

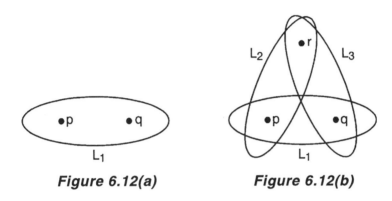

Figure 6.12(a)   **Figure 6.12(b)**

We can get other globs into play by introducing zogs and then appealing to P2 (see Figure 6.12(b)).

We are now ready to write down a formal proof.

Theorem 2. Glob Theory contains at least three globs.

Proof.

Statement	Justification
1. Let L_1 denote a glob.	1. P4
2. L_1 contains at least two zogs, p and q.	2. P1
3. There is a zog r not in L_1.	3. P3
4. There is a glob $L_2 \neq L_1$ which contains zogs p and r.	4. P2; $L_2 \neq L_1$ since L_2 contains r which is not in L_1.

Now we must be very careful. Figure 6.12(b) is helpful in providing us with a start and initial direction, but it is also potentially misleading. The suggestion conveyed is that we follow up by asserting the existence of glob $L_3 \neq L_1$ which contains r and q and exhibiting globs L_1, L_2, and L_3 to conclude the proof. But how can we be sure that L_3 is not the same as L_2? We can get around this difficulty by introducing L_3 as containing r and q and considering two cases.

5.   Case 1. L_3 is not the same as L_1 and L_2, in which case our work is done.

6.   Case 2. L_3 is the same as L_2, in which case p, q, and r are contained by the same glob, $L_3 = L_2$, visually illustrated by Figure 6.13(a).

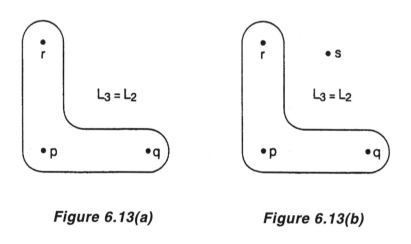

Figure 6.13(a)   **Figure 6.13(b)**

Then, by P3, there is a zog s not contained by $L_3 = L_2$, visually illustrated by Figure 6.13(b). By P2 there is a glob $L_4 \neq L_2$ which contains r and s. $L_4 \neq L_1$ contains r which is not contained by L_1. L_1, L_2, and L_4 are distinct globs.

To this point in its development Glob theory consists of two undefined terms, four postulates, and two theorems. In undertaking to prove other theorems we may now employ theorems 1 and 2 to justify assertions that we make.

The theorems, proved in the abstract for Glob Theory, hold for all models of Glob Theory well. It's just a matter of shifting from the language of zogs and globs to the terms of the model. For I_a we have:

T1a: I_a contains at least three points.

T2a: I_a contains at least three lines.

For I_b we have:

T1b: I_b contains at least three points (of the form (1,1), (5,1), (3,5)).

T2b: I_b contains at least three circular arcs.

Food for Thought

1. Specify two ideal models for Glob Theory. How do Theorems 1 and 2 of Glob Theory read in terms of these models?

2. Specify a concrete model for Glob Theory. How do Theorems 1 and 2 of Glob Theory read in terms of this model?

3. Having established that there exists at least three zogs in Glob Theory, we might attempt to climb the mathematical ladder another rung and establish that Glob Theory has at least four zogs.

   Is the following argument to prove this result valid? Explain.

   Conjecture 1: Glob Theory contains at least four zogs.

Proof.

Statement	Justification
1. Let L denote a glob.	1. P4
2. L contains at least two zogs, p and q.	2. P1
3. Let r denote a zog not on L.	3. P3
4. Let K denote a glob containing r.	4. T2

5. Let s denote another zog in K.   5. P1
6. p, q, r, and s are four zogs in   6. Summary of preceding steps.
   Glob Theory.

---

4. Michael Vlasik, an "expert" on Glob Theory, conjectured that this postulate system contains at least six zogs. Is the following argument that Michael developed to prove this conjecture valid? Explain.

Conjecture 2: Glob Theory contains at least six zogs.

Proof.

Statement	Justification
1. Let L, K, and M denote globs.	1. T2
2. Let a and b denote zogs in L, c and d zogs in K, and e and f zogs in M.	2. P1
3. a, b, c, d, e, and f are six zogs in Glob Theory.	3. Summary of preceding steps.

5. Jane Hansen, another "expert" on Glob Theory, proposed the following proof of the conjecture that this postulate system contains at least six globs. Is her argument valid? Explain.

Conjecture 3: Glob Theory contains at least six globs.

Proof.

Statement	Justification
1. Let p, q, r, and s denote four zogs.	1. Conjecture 1
2. Let G_1 denote the glob containing p and q, G_2 the glob containing p and r, G_3 the glob containing p and s, G_4 the glob containing q and r, G_5 the glob containing q and s, and G_6 the glob containing r and s.	2. P2
3. G_1 through G_6 number six globs.	3. Summary of preceding steps.

6. William Schneider, another student of Glob Theory, argued that since the proof given in 3 to prove that Glob Theory contains at least four zogs was invalid, this conjecture is not a theorem of Glob Theory. Would you agree? Explain.

7. Melinda Hu, another student of Glob Theory, expressed the view that Conjectures 1, 2 and 3 are not theorems of this system. Is she justified in this view? Explain.

8. Consider the following interpretation I_c of Glob Theory. Zog: an ordered pair of numbers which is a solution common to at least two of the following four equations, which we interpret as globs.

$$-x + y = 2, \quad -x + y = -2$$
$$x + y = 2 \qquad x + y = -2$$

   a. What are the zogs in I_c?

   b. Is I_c a model of Glob Theory? Explain.

9. **Yuk Theory**, with undefined terms luk and yuk, is based on the following five postulates.

   P1. Every yuk is a collection of luks containing at least one luk.

   P2. There are at least two luks.

   P3. For any two luks, there is one and only one yuk containing them.

   P4. For any yuk there is a luk not contained by it.

   D1. Two yuks are called **parallel** if there is no luk contained by both.

   P5. If L is a yuk and p is a luk not contained by it, there is exactly one yuk containing p which is parallel to L.

a. State two models for Yuk Theory.

The conjectures stated in (b) through (h) have been proposed concerning Yuk Theory. Consider each and prove it a theorem or show that it is not a theorem of Yuk Theory.

b. C1: Every luk is contained by at least two yuks.

c. C2: Every yuk contains at least two luks.

d. C3: There are at least three luks.

e. C4: There are at least four luks.

f. C5: There are at least four yuks.

g. C6: There are at least six yuks.

h. C7: There are at least seven yuks.

10. Consider the surface of a sphere in Euclidean space (Figure 6.14).

a. Concerning Yuk Theory, interpret luk as meaning a point on the surface of this sphere and yuk as meaning a great circle on the surface of this sphere (that is, a circle which breaks the sphere into hemispheres). Is this interpretation a model of Yuk Theory? Explain.

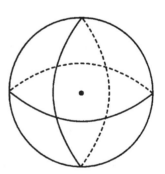

Figure 6.14

b. Concerning Glob Theory (p. 97), is this interpretation a model of Glob Theory? Explain.

11. **Neighborhood Theory**, with undefined terms blob and neighbor-hood, is based on the following three postulates.

> P1:   There are at least three blobs.
>
> P2:   For any two blobs there is at least one neighborhood containing them.
>
> P3:   For any two blobs p and q there is a neighborhood P containing p and a neighborhood Q containing q where P and Q have no blobs in common.

a.   State two models for Neighborhood Theory. The conjectures stated in (b) through (d) have been proposed concerning Neighborhood Theory. Consider each and prove it a theorem or show that it is not a theorem of Neighborhood Theory.

b.   C1: There are at least three neighborhoods.

c.   C2: Not all blobs are contained in any one neighborhood.

d.   C3: No two neighborhoods have a blob in common.

12. **Mumbo-Jumbo Theory**, with undefined terms mumbo and jumbo, is based on the following five postulates.

> P1:   For any two jumbos there is exactly one mumbo containing them.
>
> P2:   Not all jumbos are contained by the same mumbo.
>
> P3:   There is at least one mumbo.
>
> P4:   Every mumbo contains at least three jumbos.
>
> P5:   For any jumbo, there are exactly two mumbos containing it.

a.   Try to construct a model for Mumbo-Jumbo Theory. What seems to be the difficulty in doing this?

The conjectures stated in (b) and (c) have been proposed concerning Mumbo-Jumbo Theory. Consider each and prove it a theorem or show that it is not a theorem of Mumbo-Jumbo Theory.

b. C1: There are at least four jumbos.

c. C2: There is a jumbo which is contained by three mumbos.

d. If C2 is a theorem, what are its implications for Mumbo-Jumbo Theory?

6.4 CONSISTENCY AND INDEPENDENCE

A postulate system is said to be **consistent** if it does not contain contradictory statements (two postulates, postulate and theorem, or two theorems). This is the most fundamental property required of a postulate system. A postulate system that is not consistent, called **inconsistent**, is worthless. Mumbo-Jumbo Theory, stated in Exercise 12 in the previous section, is an example of an inconsistent postulate system. Conjecture C2, There is a jumbo which is contained by three mumbos, is a theorem of Mumbo-Jumbo Theory which contradicts postulate P5: For any jumbo there are exactly two mumbos containing it.

The consistency of a postulate system may take two forms, absolute and relative. The **absolute consistency** of a postulate system is established by providing a concrete model for it, which establishes that the system is as consistent as the real world. The **relative consistency** of a postulate system G is established by providing an ideal model I for it, which establishes that G is as consistent as I. Providing an ideal model I for G shifts the burden of the consistency of G to I. It shows that if G is inconsistent, then so is I.

To illustrate these ideas we return to Glob Theory, considered in the previous section, based on the following postulates.

P1:   Every glob is a collection of zogs which contains at least two zogs.

P2:   For any two zogs there is at least one glob containing them.

P3:   For any glob there is a zog not contained by it.

P4:   There is at least one glob.

Interpretations I_a and I_b discussed in the previous section are ideal models of Glob Theory which show that Glob Theory is as consistent as Euclidean geometry and the real number system, respectively. A concrete model I_d of Glob Theory is suggested by Figure 6.12(b) itself, which is reproduced in Figure 6.15.

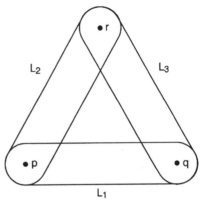

Figure 6.15

Let p, q, and r denote any three concrete objects (3 books; 3 bees in a hive; 3 teachers in a department; whatever); we take these as our zogs. For our globs we take the subsets of two of p, q, and r. We have:

$$L_1 = \{p,q\}, \quad L_2 = \{p,r\}, \quad L_3 = \{q,r\}$$

The postulates of Glob Theory take the following form in terms of this interpretation of zog and glob:

P1c: Every set of two objects of the three contains at least two objects.

P2c: For any two objects there is at least one set of two containing them.

P3c: For any set of two objects there is an object not contained by it.

P4c: There is at least one set of two objects.

These conditions are satisfied and I_d is a concrete model of Glob Theory. This establishes the absolute consistency of Glob Theory.

The idea of relative consistency was encountered in Chapter 17 (The Mathematics of Space) of *Get a Grip on Your Math* where it was noted that Eugenio Beltrami showed that Lobachevskian geometry is as mathematically respectable as Euclidean geometry by providing an ideal model for Lobachevskian geometry in Euclidean terms. To take this a step further, let us observe that analytic geometry sets up an ideal model for Euclidean geometry which establishes that Euclidean geometry is as consistent as the real number system. Thus, Lobachevskian geometry is as consistent as the real number system. By establishing ideal models the consistency of many branches of mathematics can be reduced to the consistency of a basic branch.

A postulate of a postulate system is said to be **independent** in the system if it is not a valid consequence of the other postulates of the system. If one desires a statement P within a postulate system (either as a postulate or theorem) and P is independent in the system, then the only way to have it in the system is to take it as a postulate since it cannot be had as a theorem.

One approach to showing that a postulate P is independent in a postulate system G is to exhibit an interpretation for the terms of G for which the other postulates of G are satisfied, but P is not satisfied. Having done this, it follows that P could not be a theorem of G since, if it were, if would also have to be a theorem in the aforenoted interpretation, which it is not.

Example 1

Postulate P3 of Glob Theory, based on the following postulates, is independent.

P1:   Every glob is a collection of zogs which contains at least two zogs.

P2:   For any two zogs there is at least one glob containing them.

P3:   For any glob there is a zog not contained by it.

P4:   There is at least one glob.

Consider two books b_1 and b_2. We take these as our zogs and the set $L_1 = \{b_1, b_2\}$ as our glob. Postulates P1, P2, and P4 are satisfied for this interpretation, but P3 is not. Thus, P3 is independent; if it were a theorem of postulates P1, P2, and P4, then the statement, for any set of two books, there is a book not contained by it, would be satisfied in this interpretation, which is not the case.

As discussed in Chapter 17 (The Mathematics of Space) of *Get a Grip on Your Math*, the most famous consideration of the independence of a postulate arose in connection with Euclid's parallel postulate. It was firmly believed for 2200 years that Euclid's parallel postulate was not independent of his other postulates. The development of a non-Euclidean geometry by Lobachevsky based on a contradiction to Euclid's parallel postulate and the proof of its relative consistency in terms of Euclidean geometry not only showed the independence of Euclid's parallel postulate, but was responsible for shaping attitudes which underlie the nature of modern mathematics.

Food for Thought

1. Victor P. Neighborhood, son of the founder of Neighborhood Theory (Sec. 6.3, Ex. 11, p. 107) proposed an extension of Neighborhood Theory by adding the following postulate to P1–P3. P4: There exist at most two neighborhoods. Is Victor's Extended Neighborhood Theory consistent?

2. Janice Neighborhood, on the other hand, proposed an extension by adding the following postulate to P1–P3. P4(a): There exist at most three neighborhoods.

   a. Is Janice's Extended Neighborhood Theory consistent? Explain.

   b. Is P4(a) independent? Explain.

3. Is Yuk Theory, defined in Question 9 of Section 6.3 (p. 105), consistent? Explain.

4. Marcel Yuk formulated modified Yuk Theory (MYT) by adding the following postulate P6 to the five postulates of Yuk Theory. P6: There are at most five yuks. Is MYT consistent? Explain.

5. Is postulate P5 of Yuk Theory independent?

6. An **Abelian Group** is a postulate system which consists of a set G of objects, called **elements of G**, and an operation denoted by ·, subject to the following postulates:

   P1:   To every pair of elements x and y of G, given in the stated order there corresponds an unique element of G, denoted by x · y. (Closure postulate for the operation ·.)

   P2:   If x, y, and z are any elements of G, then x · (y · z) = (x · y) · z. (Associative postulate for the operation ·.)

   P3:   There is an unique element e of G, called the identity element, having the property that if x is any element of G, x · e = e · x = x.

   P4:   To each element x of G there corresponds an unique element x′, called the inverse of x, having the property that x · x′ = x′ · x = e.

   P5:   If x and y are any elements of G, then x · y = y · x. (Commutative postulate for the operation ·.)

   a.   Is Group Theory consistent? Explain.

   b.   Show that postulate P5 is independent.

6.5 MATHEMATICAL MODELS FOR REAL WORLD PHENOMENA

The term **mathematical model**, often shortened to **model**, is also used in connection with developing a mathematical portrait of a real world phenomenon of interest. Just as a person's portrait may be sketched in many ways, so too may a mathematical portrait of a real world phenomenon be developed in many ways, depending on its features of interest to be captured and the ingenuity and sensitivity of the mathematical artist developing the model. A mathematical portrait consists of assumptions, or equivalently, postulates, that the mathematical artist is willing to make, postulates which lead to valid conclusions—the theorems of the model.

How do we determine how "successful" is a mathematical portrait of our subject of interest? If its postulates are "reasonably" accurate, as determined by experimentation and observation, then the model is judged a success. The problem is that in general a model's postulates do not lend themselves to a truth/falsity determination by experimentation and observation. This being the case, close attention must be paid to the realism of the postulates of the model. In this situation we are like the artist sketching a portrait in dim light. We would like to turn a bright light on the subject, but it is not available and so we must proceed carefully with the dim light we do have.

Another source of light is to be found in the model's theorems. If their truth/falsity can be established, then doing so will throw light, indirect light, to be sure, on the postulates. If the theorems are shown to be true, or to allow for an important shade of gray, realistic, then the postulates, as we saw illustrated in Section 6.1, may be realistic or possibly unrealistic. Realistic theorems may arise from unrealistic postulates, so that establishing the realism of a model's theorems does not definitively establish the realism of its postulates. But it does serve to strengthen our confidence in the postulates as providing a good working portrait of the phenomenon under study, with the understood qualification that there may be important aspects of the portrait which are incomplete and possibly inaccurate. On the other hand, if a theorem of the model is show to be false, then this sends us an unequivocal message that the portrait provided by its postulates is in need of revision. The revision called for may be relatively minor or very drastic. Chapter 8 takes up illustrations of this situation in considering a variety of probability models for "simple" random processes.

Suppose, for example, that a production model has been developed for a company that manufactures stereo systems, product models RA-5 and RA-9, let us say. A theorem of this model says that for the company to maximize its monthly profit on RA-5 and RA-9 it should set its production schedule to make 10,000 RA-5 and 12,000 RA-9 units per month for a projected maximum profit of $750,000. This valid conclusion of the model may sound good, but is it realistic? The acid test entails setting the production schedule at these levels and seeing if the actual monthly profit realized is in the neighborhood of the projected $750,000. If it turns out not to be the case, that is, we have a valid but unrealistic conclusion, this acid test could prove rather costly to the company. The question is, what options are available prior to going ahead with the acid test and implementing the derived production schedule? Only one, and that is to pay particularly close attention to the realism of the postulates of the model. This is the decisive juncture. If there are reservations about the realism of the assumptions being made, then such should be resolved before implementing the production schedule obtained as a valid conclusion. This may require that the model, that is, its postulates, be refined and that a new production schedule be derived from the refined model. Illustrations of this situation are taken up in Chapter 7.

Food for Thought

1. Consider the following statements; for each one state, with explanation, whether you agree or disagree.

    a. If some of a model's theorems have been confirmed by experimentation/observation, then the model's postulates must be realistic.

    b. If some of a model's postulates are unrealistic, then some of its theorems must be false.

    c. If a model's postulates are unrealistic, then some of its theorems may be true.

    d. If a model's postulates are true, then some of its theorems may be false.

    e. Postulates of a model, by their very meaning, are true statements.

    f. If a theorem is false, then it cannot be a valid statement.

    g. If a conclusion obtained from a model is true, then it must be a theorem of the model.

h. A conclusion obtained from a model can be shown to be a theorem by showing that it is true.

2. Raymond J. Ellis of Ecap University developed a mathematical model to describe the nature and behavior of gravity waves. The realism of the postulates of Ellis's model cannot be determined by directly subjecting them to experimental verification.

   a. What can be done to access the realism of Ellis's model? Explain.

   b. Over a period of twenty years ten valid conclusions of Ellis's model were subjected to experimental tests.

      i. Suppose all ten of these conclusions were shown to be realistic. Would this establish that Ellis's model is as close to a perfect portrait of gravity waves that one could hope to obtain? Explain.

      ii. Suppose the first nine of Ellis's conclusions were shown to be realistic, but that the tenth was shown to be unrealistic. What implications would this have for the Ellis model?

3. A team of economists headed by Janet Valdez developed a mathematical model for foreign trade between a group of countries in Central and South America. One of the valid consequences of this model holds that the gross domestic product of each of the participating countries would rise by at least 5% per year for ten years if all tariffs between the countries were eliminated. Before implementing this conclusion by eliminating tariffs a commission made up of the economic ministers of the countries involved undertook to review the Valdez model. What should be the focus of the review? How so?

4. Jupiter Motors is in the enviable position of having a shipping schedule problem. Its Jupiter 500 sport car is very popular and it has become difficult to keep up with demand. J.M. is obligated to supply its distributors in New York, Atlanta, Dallas, and Seattle 6100, 8200, 6500, and 9000 Jupiter 500's per month, respectively. The cars are to be shipped from plants in Arkin, Hastings, and Freelawn. The problem is to satisfy these commitments at minimum total cost.

Operations research groups, working independently, set up two mathematical shipping models, designated by SM-1 and SM-2, for this problem. A valid consequence of SM-1 is that the shipping schedule shown in Table 6.1 should be implemented to obtain a minimum monthly shipping cost of $650,000. A valid consequence of SM-2 is that the shipping schedule shown in Table 6.2 should be implemented to obtain a minimum monthly shipping cost of $725,000.

Table 6.1 Shipping Schedule Based on SM-1

	New York	Atlanta	Dallas	Seattle
Arkin	2000	4500	3000	1500
Hastings	1100	2500	1500	3000
Freelawn	3000	1200	2000	4500

Table 6.2 Shipping Schedule Based on SM-2

	New York	Atlanta	Dallas	Seattle
Arkin	1000	3000	4000	3000
Hastings	3200	1500	1000	2400
Freelawn	1900	3700	1500	3600

a. Would you implement SM-1's shipping schedule because its projected total minimum cost, $650,000 per month, is less than the $725,000 projected total minimum cost of implementing SM-2? Explain.

b. If your answer to (a) is no, on what basis would you implement one of the aforenoted shipping schedules?

c. Is it possible that you might not implement either of the aforenoted shipping schedules? How so?

d. Does the lower projected cost of SM-1 imply that SM-2's projected cost is not valid? How so?

LINEAR PROGRAM MODELS

7.1 RETURN TO THE AUSTIN COMPANY

The Austin Company, let us recall from *Get a Grip on Your Math* (Ch. 11; A Tale of Two Linear Programs), has decided to enter the digital tape player market by introducing two models, DT-1 and DT-2, into the market. Their problem is to determine the number of units of each model to be produced to maximize profit.

The Company's operations research department was asked to study the situation and make recommendations. The OR department began their analysis by collecting data. They divided the manufacturing process into three phases; construction, assembly, and finishing. The data collected and their analysis led them to make the following assumptions.

a. In the construction phase each DT-1 unit requires 2 hours of labor and each DT-2 unit requires 3 hours of labor. At most 1,100 hours of construction time are available per week.

b. In the assembly phase each DT-1 unit requires 5 hours of labor and each DT-2 unit requires 3 hours of labor. At most 1,400 hours of assembly time are available per week.

c. In the finishing phase each DT-1 unit requires 4 hours of labor and each DT-2 unit requires 1 hour of labor. At most 756 hours of finishing time are available per week.

d. After taking cost and revenue factors into consideration the anticipated profit for each DT-1 unit is $150 and the anticipated profit for each DT-2 unit is $120. In order for these unit profit values to hold the Company must produce at least 25 DT-1 and 40 DT-2 units per week.

e. There is an unlimited market for the DT-1 and DT-2 models.

f. All factors other than the ones considered in the analysis of the production of the DT-1 and DT-2 models are negligible.

Its next task was to translate these assumptions into mathematical form, being careful to include everything stated in the assumptions and not go beyond them. The OR department began by introducing variables for the quantities it sought to determine; let x denote the number of DT-1 and y the number of DT-2 units to be made weekly. There is a fair amount of data contained in the assumptions and it is useful to make it available at a glance in tabular form. This is done in Table 7.1.

Table 7.1

No. of Units to be made		Profit per unit	Construction time per unit (hrs)	Assembly time per unit (hrs)	Finishing time per unit (hrs)
DT-1	x	$150	2	5	4
DT-2	y	$120	3	3	1

The key to expressing profit and the conditions that have emerged in terms of x and y is the information stated on unit profits and unit construction, assembly and finishing times for DT-1 and DT-2. We have:

$$\text{profit} = \begin{bmatrix} \text{profit on} \\ \text{DT-1} \end{bmatrix} + \begin{bmatrix} \text{profit on} \\ \text{DT-2} \end{bmatrix}$$

$$= \begin{bmatrix} \text{profit on} \\ \text{one DT-1} \\ \text{unit} \end{bmatrix} \cdot \begin{bmatrix} \text{no. of} \\ \text{units} \\ \text{made} \end{bmatrix} + \begin{bmatrix} \text{profit on} \\ \text{one DT-2} \\ \text{unit} \end{bmatrix} \cdot \begin{bmatrix} \text{no. of} \\ \text{units} \\ \text{made} \end{bmatrix}$$

$$= 150x + 120y$$

The profit obtained by making x DT-1 and y DT-2 units per week is expressed by the linear function:

$$P(x,y) = 150x + 120y$$

As to the conditions that x and y must satisfy, since the number of units made must be non-negative, we have:

$$x \geq 0$$

$$y \geq 0$$

The construction time condition is that

$$(\text{total construction time used}) \leq 1100.$$

In terms of unit construction times, 2 hours are needed for one unit of DT-1 and 3 hours are needed for one unit of DT-2, 2x hours are needed for x DT-1 units and 3y hours are needed for y DT = 2 units. The total construction time utilized is expressed by 2x + 3y. Thus, the construction time utilized is expressed by 2x + 3y. Thus, the construction time condition is:

$$2x + 3y \leq 1100$$

Similarly, the assembly and finishing time conditions are stated by the inequalities:

$$5x + 3y \leq 1400$$

$$4x + y \leq 756$$

The conditions that at least 25 DT-1 and 40 DT-2 units must be produced weekly are expressed by the inequalities:

$$x \geq 25$$

$$y \geq 40$$

We thus emerge with the following mathematical structure, called linear program model LP-1, as a translation of the assumptions made by the OR department of the Austin Company.

$$\text{Maximize } P(x,y) = 150x + 120y$$

subject to

$$x \geq 0, y \geq 0$$

$$2x + 3y \leq 1100$$

$$5x + 3y \leq 1400$$

$$4x + y \leq 756$$

$$x \geq 25, y \geq 40$$

Here x represents the number of DT-1 and y the number of DT-2 units to be made weekly.

More generally, a **linear program** is a mathematical problem with the following structure: There is specified a linear function of a number of variables that are required to satisfy linear conditions described by some mixture of linear inequalities and linear equations, called **constraints**. The problem is to find values for these variables which satisfy the constraints and yield the maximum, or minimum, value of the function, which is called an **objective function**. LP-1 is a 2-variable linear program; the problem of finding values of $x_1, x_2, \ldots, x_{20}$ that satisfy the constraints

$$x_1 \geq 0, \; x_2 \geq 0, \ldots, x_{20} \geq 0$$

$$x_1 - \; x_2 + \ldots - \; x_{20} \leq 20$$

$$x_1 + 2x_2 + \ldots + 20x_{20} \leq 200$$

and maximize the objective function $G(x_1, \ldots, x_{20}) = 2x_1 - 4x_2 + \ldots + 38x_{19} - 40x_{20}$ is a 20-variable linear program. The non-negativity constraints, $x \geq 0$, $y \geq 0$ in the first example, and $x_1 \geq 0, \ldots, x_{20} \geq 0$ in the second example, are present in many linear programs, but not all. The functions and constraints are linear in that all variables occur to the first power and there are no products of variables.

The Austin Company, as we noted in *Get a Grip on Your Math*, also hired the Marks Company, a consulting operations research firm, to independently study the digital tape player situation and make recommendations. The Marks OR group divided the manufacturing process into two phases: construction (which included assembly) and finishing. The data collected and their analysis led them to make the following assumptions:

a. In the construction phase each DT-1 unit requires 8 hours of labor and each DT-2 unit requires 5 hours of labor. At most 2,210 hours of construction time are available per week.

b. In the finishing phase each DT-1 unit requires 3 hours of labor and each DT-2 unit requires 2 hours of labor. At most 860 hours of finishing time are available per week.

c. The anticipated profit for each DT-1 unit is $140 and the anticipated profit for each DT-2 unit is $150. In order for these unit profit values to hold the company must produce at least 50 DT-1 and 50 DT-2 units per week.

d. There is an unlimited market for the DT-1 and DT-2 models.

e. All factors other than the ones considered in the analysis of the production of the DT-1 and DT-2 models are negligible.

The same sort of analysis that leads to LP-1 from the assumptions made by the Austin Company's operation research department leads to the Marks OR group's linear program model LP-2:

$$\text{Maximize } P(x,y) = 140x + 150y$$

subject to

$$x \geq 0, \, y \geq 0$$
$$8x + 5y \leq 2210$$
$$3x + 2y \leq 860$$
$$x \geq 50, \, y \geq 50,$$

where x represents the number of DT-1 and y the number of DT-2 units to be made weekly.

How can such problems be solved? We develop some needed preliminaries in the next section and the corner point solution method for 2-variable linear programs in Section 7.3.

7.2 LINEAR STRUCTURES

SYSTEMS OF TWO LINEAR EQUATIONS IN TWO UNKNOWNS

The problem of solving systems of two linear equations in two unknowns arises in the application of the corner point method. We begin by turning our attention to this important preliminary problem.

Example 1

Solve the system of linear equations:

$$3x + y = 9$$
$$x + 2y = 8$$

By a **solution** to such a system we mean any ordered pair of numbers, one for x and the other for y, that satisfies both equations in the system. Thus, x = 3, y = 0, also denoted by (3,0), is not a solution of this system since substitution of 3 for x and 0 for y yields:

$$3(3) + 0 = 9$$
$$3 + 2(0) = 8$$

The first of these statements holds, but clearly the second does not.

Geometrically speaking, linear equations in two variables translate to lines and a pair of such equations translates to a pair of lines. The question of how many solutions a pair of linear equations in two unknowns can have translates geometrically to the question of how does a pair of lines (in the same coordinate plane) behave. The answer, of course, is that two lines may intersect in one point (distinct lines), not intersect (parallel lines), or completely overlap (identical lines). In terms of solutions this means that a system of two linear equations in two unknowns may have one solution (described by the single point of intersection of their graphs), no solution, or an uncountable number of solutions (we say an infinite number of solutions). Our interest is in the first of these situations.

To find solutions of systems of equations we shall employ, in a systematic way, techniques that change the appearance of the equations of a system, but do not alter the solutions of the system. Consider the following principles.

P1.   If both sides of an equation are multiplied by a nonzero constant, then the resulting equation has the same solutions as the original one.

For example, $3x + y = 9$ and $-2(3x + y = 9)$, that is, $-6x - 2y = -18$, have the same solutions.

P2.   If two equations have a common solution, then this common solution is also a solution of the sum of the two equations.

For example, the equations $3x + y = 9$ and $x + 2y = 8$ have $(2,3)$ as a solution in common. Their sum, $4x + 3y = 17$, also has $(2,3)$ as a solution.

By combining principles P1 and P2 we obtain the following:

P3.   If two equations have a common solution, then this common solution is also a solution of the sum of one of the equations and a nonzero constant multiple of the other.

For example, the equations $3x + y = 9$ and $x + 2y = 8$ have $(2,3)$ as a solution in common. The sum of $-2(3x + y = 9)$, that is, $-6x - 2y = -18$, and $x + 2y = 8$, which is $-5x + 0y = -10$, also has $(2,3)$ as a solution.

These principles lead to the following three rules of operation which can be used to solve systems of linear equations.

RULES OF OPERATION

R1.   An equation in a system of linear equations may be multiplied by a nonzero constant.

R2. Any equation in a system of linear equations may be added to any other equation in the system.

R3. A nonzero constant multiple of any equation in a system of linear equations may be added to any other equation in the system.

With these rules of operation in place we return to the system to be solved:

$$3x + y = 9 \tag{1}$$

$$x + 2y = 8 \tag{2}$$

The problem is that there are two unknowns to contend with, the mathematical equivalent of a two front war. Our strategy, then, is to take one of the unknowns out of scene, deal with the other, and return to the first. The easiest way to proceed is to multiply (1) by –2 and add the result to (2), thus taking y out of the scene.

In summary we have:

$$-6x - 2y = -18 \tag{3}$$

$$\underline{x + 2y = 8} \tag{2}$$

$$-5x = -10$$

$$x = 2$$

To determine y we replace x by 2 in one of the equations, (2) for example, and solve for x. We have:

$$x + 2y = 8$$
$$2 + 2y = 8$$
$$2y = 6$$
$$y = 3$$

What we have shown is that if our system has a solution, then (2,3) is it; it's the only possibility.

To nail this down and establish that (2,3) is a solution to our system we must replace x by 2 and y by 3 in (1) and (2) and verify that both conditions are satisfied. We have

$$3(2) + 3 = 9$$
$$2 + 2(3) = 8,$$

both of which are correct.

Example 2

$$\text{Solve:} \quad 5x + 3y = 1400 \tag{4}$$
$$4x + y = 756 \tag{5}$$

The easiest way to proceed is to take y out of the scene by multiplying (5) by −3 and adding (4) to the result. We obtain:

$$5x + 3y = 1400 \tag{4}$$
$$\underline{-12x - 3y = -2268} \tag{6}$$
$$-7x = -868$$
$$x = 124$$

Replacing x by 124 in (5) and solving for y yields:

$$4(124) + y = 756$$
$$y = 260$$

We verify that (124,260) is the solution of our system by replacing x by 124 and y by 260 in (4) and (5) and noting that correct numerical statements are obtained in each case.

Food for Thought

Solve the following systems of linear equations. As we shall see, these problems arise in connection with solving linear programming problems in Section 7.3.

1. $2x + 3y = 1100$
   $x = 25$

2. $4x + y = 756$
   $y = 40$

3. $3x + 2y = 860$
   $x = 50$

4. $8x + 5y = 2210$
   $y = 50$

5. $2x + 3y = 15$
   $x + y = 6$

6. $3x + 5y = 12$
   $x + y = 3$

7. $2x + y = 4$
   $x + y = 3$

8. $x + y = 3$
   $x + 2y = 4$

9. $x + y = 60$
   $-x + 3y = 0$

10. $x + y = 250{,}000$
    $2x + y = 400{,}000$

11. $5x + 3y = 1400$
    $2x + 3y = 1100$

12. $3x + 2y = 860$
    $8x + 5y = 2210$

13. $4x + 3y = 320$
    $5x + 2y = 330$

14. $5x + 2y = 330$
    $3.25x + 2y = 225$

15. $4x + 3y = 320$
    $3.25x + 2y = 225$

LINEAR INEQUALITIES IN TWO VARIABLES

A linear equation in two variables, x and y, let us say, is an equation which can be expressed in the form $Ax + By = C$, where A, B and C are constants with not both A and B being O. The equations that we encountered in the preceding are linear equations in two variables (or unknowns) since they all have this form.

The graph of the general 2-variable linear equation $Ax + By = C$ is a **boundary line** which divides the rest of the coordinate plane into two parts. On one side of $Ax + By = C$ the points satisfy $Ax + By < C$ while on the other side they satisfy $Ax + By > C$. To apply this to the constraint $x + 2y \leq 8$, for example, we first plot the line $x + 2y = 8$ (see Figure 7.1), which serves as a boundary line. A systematic procedure for determining which side of the boundary line describes x + 2y < 8 and which side describes x + 2y > 8 involves taking a test point not on the boundary line and determining which inequality condition it satisfies. When the boundary line does not pass through the origin, (0,0) is usually a convenient choice, but any point not on the boundary line will do. Since (0,0) satisfies the constraint x + 2y < 8, the graph of x + 2y ≤ 8 consists of all points on the boundary line x + 2y = 8 together with those points on the same side of the boundary line as our test point (0,0), that is, below the boundary line. This is indicated by lining-in the region below the boundary line (see Figure 7.2).

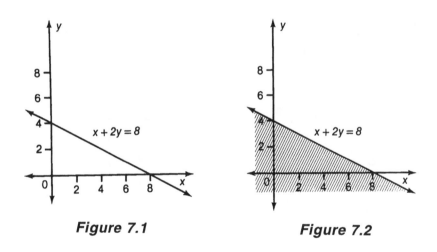

Figure 7.1   **Figure 7.2**

Example 3

Sketch the graph of the system:

$$x + 2y \leq 8$$
$$3x + y \leq 9$$

To graph a system of inequalities we graph the inequalities that make up the system on the same coordinate system and pick out the region which is common to them all.

In this case, we begin by graphing $x + 2y \leq 8$ as just discussed. Next we graph $3x + y \leq 9$ on the same coordinate system. This yields Figure 7.3. The points which satisfy both of the constraints $x + 2y \leq 8$ and $3x + y \leq 9$ lie on the intersection of their graphs, the region below both boundary lines and the adjoining parts of these boundary lines, shown in Figure 7.4.

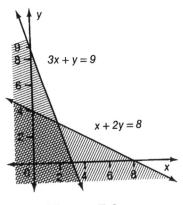

Figure 7.3

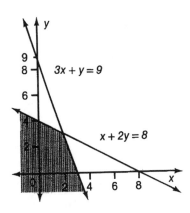

Figure 7.4

Example 4

Sketch the graph of the system:

$$x \geq 0, y \geq 0$$
$$x + 2y \leq 8$$
$$3x + y \leq 9$$

The non-negativity constraints $x \geq 0$, $y \geq 0$ restrict us to the first quadrant. We thus obtain Figure 7.5, the graph of Figure 7.4 restricted to the first quadrant.

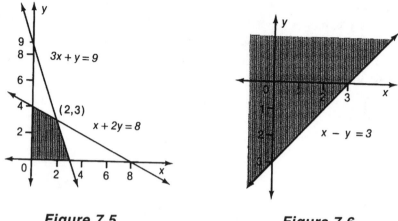

Figure 7.5   **Figure 7.6**

In graphing constraints of the form Ax + By ≤ C or Ax + By ≥ C, the test point device to determine which side of the boundary line Ax + By = C is a routine which always works. Some simplification, however, is noteworthy. If the coefficients A and B are both positive, then the ≤ condition yields all points on the boundary line and below it; the ≥ condition yields all points on the boundary line and above it. If one or both of the coefficients A and B are negative, then play it safe by using the test point device; anything may happen and it does not pay to consider the various special cases that may arise.

Example 5

Sketch the graph of x − y ≤ 3.

We first plot the boundary line x − y = 3 and then choose a test point not on this line; (0,0) will do as a test point, and we see that it satisfies x − y ≤ 3. Since this test point is above x − y = 3, the graph of x − y ≤ 3 consists of all points on the boundary line x − y = 3 and above it, even though we have a less than or equal to (≤) inequality; see Figure 7.6.

Example 6

Sketch the graph of the constraints of LP-1, the linear program model obtained by the operations research department of the Austin company.

$$x \geq 0, y \geq 0$$
$$2x + 3y \leq 1100$$
$$5x + 3y \leq 1400$$
$$4x + y \leq 756$$
$$x \geq 25, y \geq 40$$

By proceeding systematically and graphing each constraint, restricting ourselves to the first quadrant ($x \geq 0$, $y \geq 0$), we obtain the graph shown in Figure 7.7.

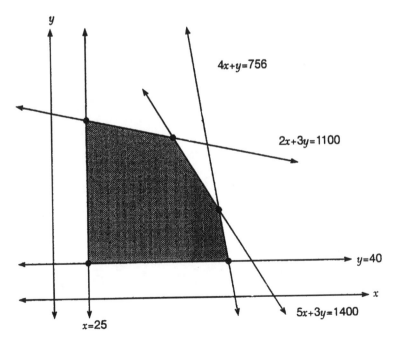

Figure 7.7

Food for Thought

Sketch the graphs of the following systems of constraints. As we shall see, most of these systems arise in connection with solving linear programming problems in Section 7.3.

16. $x \geq 0, y \geq 0$
$2x + 3y \leq 15$

17. $x \geq 0, y \geq 0$
$2x + 3y \leq 15$
$x + y \leq 6$

18. $x \geq 0, y \geq 0$
$3x + 5y \leq 12$

19. $x \geq 0, y \geq 0$
$3x + 5y \leq 12$
$x + y \geq 3$

20. $x \geq 0, y \geq 0$
$2x + y \geq 4$
$x + y \geq 3$

21. $x \geq 0, y \geq 0$
$2x + y \geq 4$
$x + y \geq 3$
$x + 2y \geq 4$

22. $x \geq 0, y \geq 0$
$x + y \geq 250,000$
$2x + y \leq 400,000$

23. $x \geq 0, y \geq 0$
$x + y \leq 60$
$-x + 3y \geq 0$
$x \geq 15$

24. $x \geq 0, y \geq 0$
$x + y \leq 100$
$x + y \geq 15$
$x \leq 75, y \leq 90$

25. $x \geq 0, y \geq 0$
$8x + 5y \leq 2210$
$3x + 2y \leq 860$
$x \geq 50, y \geq 50$

26. $x \geq 0, y \geq 0$
$4x + 3y \leq 320$
$5x + 2y \leq 330$

27. $x \geq 0, y \geq 0$
$5x + 2y \leq 330$
$3.25x + 2y \leq 225$
$4x + 3y \leq 320$

7.3 THE CORNER POINT METHOD

How are LP-1 and LP-2, the linear program models that arose in connection with the Austin Company's production scheduling problem, to be solved? To do this we develop a simple method for solving linear programs called the **corner point method**. It has an appealing geometric flavor and is effective for 2-variable linear programs.

As a working illustrative example consider the linear program:

$$\text{Maximize } F(x,y) = 5x + 8y$$

subject to

$$x \geq 0, \, y \geq 0$$
$$x + 2y \leq 8$$
$$3x + y \leq 9.$$

The points (ordered pairs of numbers in this case) which satisfy the constraints of a linear program, called **feasible points**, are the points that the objective function to be optimized may be applied to.

Our problem here is to determine that feasible point (or those feasible points, if there is more than one) which yields the maximum value for the objective function $F(x,y) = 5x + 8y$.

Our first step in solving this problem is to obtain a geometric representation of the feasible points by graphing the constraints. This is done in Example 4 of the preceding section (p. 129).

The intersection points of at least two boundary lines which come out of the equality conditions of the constraints of a 2-variable linear program are called **corner points**. The corner points of our linear program are (0,0), (0,4), (3,0) and (2,3)—obtained by solving the system of equations $x + 2y = 8$ and $3x + y = 9$ (see Sec. 7.2, Example 1, p. 123); they are shown in Figure 7.8.

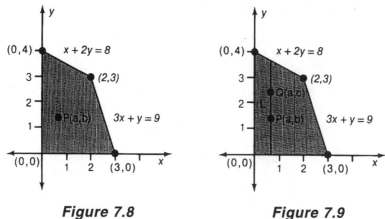

Figure 7.8   **Figure 7.9**

The significance of corner points is made clear by the following argument. As a starting point consider any feasible point P(a,b) in the region of feasible points of our linear program; see Figure 7.8. If we move up the vertical line L passing through P(a,b) to Q(a,c), (see Figure 7.9), our objective function $F(x,y) = 5x + 8y$ increases in value from $5a + 8b$ to $5a + 8c$ since c is larger than b. By taking feasible points higher and higher on L we increase $F(x,y) = 5x + 8y$. Since we must remain within the set of feasible points in taking points higher and higher on L, we can go as far as R(a,d) on the boundary line $x + 2y = 8$ (Figure 7.10). From there we can move in one of two directions on the boundary line until we come to a corner point (see Figure 7.11).

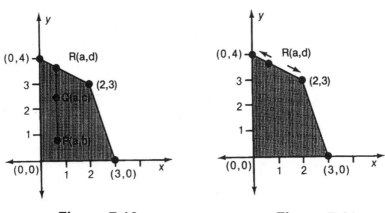

Figure 7.10   **Figure 7.11**

The question that this raises is, how does a linear function behave as we take points in one direction or the other along a boundary line? We shall not do so here, but one can prove that one of two things happens: (a) The linear function increases in value as we take points in one direction along the boundary line and decreases as we take points in the other direction, or (b) the linear function has the same value at all the points on the boundary line. In either case we are led to a corner point. This argument suggests the following theorem.

Corner Point Theorem: If a 2-variable linear program has an optimal value (maximum or minimum value, depending on the nature of the linear program), then a solution yielding this optimal value can be found from among the corner points of the linear program.

We have by no stretch of the imagination given a proof of this result. We have given a preproof argument which is intended to indicate that the result is valid and can be established by proof which conforms to rigorous mathematical standards of proof.

The corner point theorem's hypothesis presupposes that the linear program under consideration has a solution. It does not say that a solution cannot occur at a feasible point which is not a corner point; this does happen with some linear programs. We are, however, assured of a solution at a corner point, assuming that the linear program has a solution to begin with.

Implementation of the corner point method to solve a 2-variable linear program involves the following sequence of steps:

1. Graph the feasible points of the linear program.

2. Locate its corner points on the graph.

3. Determine the coordinates of all corner points. For a corner point which is not on either of the coordinate axes this is done by solving the system of equations which describe a pair of boundary lines which intersect at the corner point.

4. Compute the value of the objective function at each corner point.

5. From these values pick out the largest or smallest value, depending on the nature of the linear program, and the solution(s) which yields it.

To illustrate these procedures we return to our illustrative working example:

$$\text{Maximize } F(x,y) = 5x + 8y$$

subject to

$$x \geq 0, \, y \geq 0$$
$$x + 2y \leq 8$$
$$3x + \, y \leq 9$$

The graph of its feasible points is reproduced as Figure 7.12. The corner points, displayed in Figure 7.13, are (0,0), (3,0), (2,3) and (0,4); as was previously noted, (2,3) is obtained by solving the system of boundary line equations $3x + y = 9$ and $x + 2y = 8$.

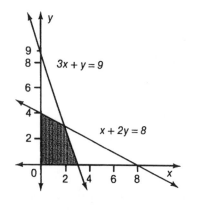

Figure 7.12

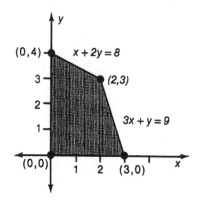

Figure 7.13

The computation of the value of the objective function F(x,y) = 5x + 8y at the corner points yields the results summarized in Table 7.2, from which we see that 34 is the maximum value and (2,3) is the solution.

Table 7.2

Corner Point	F(x,y) = 5x + 8y
(0,0)	0
(3,0)	15
(2,3)	34
(0,4)	32

Example 1

Solve LP-1, the linear program model obtained by the operations research department of the Austin Company.

Maximize P(x,y) = 150x + 120y

subject to

$x \geq 0, y \geq 0$

$2x + 3y \leq 1100$

$5x + 3y \leq 1400$

$4x + y \leq 756$

$x \geq 25, y \geq 40$

Our first step is to sketch the graph of the feasible points. This is done in Example 6 of Section 7.2 (p. 131).

Locate the corner points on the graph and solve the appropriate systems of equations to determine their coordinates. There are five corner points, shown in Figure 7.14: (25,40), (25,350), (100,300)—obtained by solving $2x + 3y = 1100$ and $5x + 3y = 1400$, (124,260)—obtained by solving $4x + y = 756$ and $5x + 3y = 1400$, and (179,40).

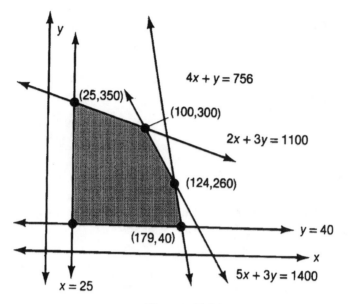

Figure 7.14

The computation of $P(x,y) = 150x + 120y$ at the corner points, summarized in Table 7.3, yields the solution (100,300) with maximum value 51,000.

Table 7.3

Corner Point	$P(x,y) = 150x + 120y$
(25,40)	8,550
(25,350)	45,750
(100,300)	51,000
(124,260)	49,800
(179,40)	31,650

WHICH SOLUTION SHOULD BE IMPLEMENTED?

This is the question facing the Austin Company. Model LP-1 has solution (100,300) with maximum value 51,000 whereas model LP-2 has solution (50,355) with maximum value 60,250 (see Exercise 9). In terms of the Austin Company's situation, to implement LP-1's conclusion the production schedule would have to be set to manufacture 100 DT-1 and 300 DT-2 units per week with an anticipated weekly profit of $51,000; to implement LP-2's conclusion the production schedule would have to be set to manufacture 50 DT-1 and 355 DT-2 units per week with an anticipated weekly profit of $60,250.

As Bottom-line Bob found, much to his surprise and distress (*Get a Grip on Your Math*, Ch. 11), it would be a serious error to implement the (50,355) production schedule with the idea that mathematical methods—the corner point method in this case—guarantee that the higher weekly profit of $60,250 will be realized. The corner point method, as a mathematical method, guarantees that both conclusions are valid with respect to the linear program models from which they arose. The question is, which of these valid conclusions is realistic? The corner point method, as a mathematical technique, provides no answer to this question. To answer this question we must go back to models LP-1 and LP-2 and examine their assumptions from the point of view of realism. Implement that valid conclusion which stems from realistic assumptions, for then its valid conclusion will be realistic. If the assumptions of both LP-1 and LP-2 are found to be unrealistic, do not implement either of the production schedules arising from these models. Another production scheduling model, hopefully with realistic assumptions, would have to be formulated.

Food for Thought

1. ZKB Electronics puts out two kinds of personal computers, model ZKB-47 and model ZKB-82. The management of ZKB called in the Vytis Consulting Firm to determine how many units of each model should be made daily so as to maximize profit. The consulting firm set up a linear program model for the electronics company's production problem and, by applying the corner-point method, reached the conclusion that 300 ZKB-47 units and 250

ZKB-82 units should be made daily to maximize profit. Before implementing this conclusion, the management of ZKB put the following questions to the director of the consulting firm.

a. Does use of the corner-point method guarantee that profit will be maximized when 300 ZKB-47 units and 250 ZKB-82 units are made daily and sold?

b. What is your basis for recommending that we implement your conclusion?

How would you reply to these questions?

2. The Onute Corporation plans to introduce two high resolution TV models, T20 and T24, to the market. Its own operations research group was led to introduce the following M1 model to determine the optimal production schedule for maximizing profit:

$$\text{Maximize } P(x,y) = 180x + 120y$$

subject to

$$x \geq 0, \, y \geq 0$$
$$4x + 3y \leq 320$$
$$5x + 2y \leq 330,$$

where x and y denote the number of T20 and T24 units, respectively, to be made daily. M1 is based on the assumptions described in Exercise 1 of Section 7.5 (p. 160). Its solution is (50,40) with maximum value 13,800 (see Exercise 10).

The Alexis company was also hired to study the Onute Corporation's production scheduling problem. Based on the assumptions described in Exercise 2 of Section 7.5 it was led to introduce the following M2 model to determine the optimal production schedule for maximizing profit:

Maximize P(x,y) = 190x + 110y

subject to

x ≥ 0, y ≥ 0

5x + 2y ≤ 330

3.25x + 2y ≤ 225

4x + 3y ≤ 320,

where x and y denote the number of T20 and T24 units, respectively, to be made daily. Its solution is (60,15) with maximum value 13,050 (see Exercise 11).

The following questions have arisen. How would you answer them?

   a. If mathematics is the precise subject that it is reputed to be, should there not be one solution to this problem rather than two?

   b. Since two solutions emerge, does it follow that not both are valid? Explain.

   c. Before making a decision about whether to implement M1 or M2, what questions would you put to the two operations research groups?

   d. Which model, if either, would you adopt and implement? Why? Is it possible that you would not adopt either model?

Solve the following linear programs by the corner point method.

3.   Maximize F(x,y) = 5x + 4y

subject to

    x ≥ 0, y ≥ 0
    2x + 3y ≤ 15
    x + y ≤ 6

4.   Minimize G(x,y) = 10x + 12y

subject to

    x ≥ 0, y ≥ 0
    3x + 5y ≤ 12
    x + y ≥ 3

5.   Minimize C(x,y) = 1.50x + 1.10y

subject to

    x ≥ 0, y ≥ 0
    2x + y ≥ 4
    x + y ≥ 3
    x + 2y ≥ 4

6.   Maximize I(x,y) = 0.10x + 0.08y

subject to

    x ≥ 0, y ≥ 0
    x + y ≤ 60
    −x + 3y ≥ 0
    x ≥ 15

7.   Minimize C(x,y) = 8x + 12y

subject to

$$x \geq 0, y \geq 0$$
$$x + y \geq 250,000$$
$$2x + y \leq 400,000$$

8.   Minimize C(x,y) = −5x + 12,795

subject to

$$x \geq 0, y \geq 0$$
$$x + y \leq 100$$
$$x + y \geq \ \ 15$$
$$x \leq 75, y \leq 90$$

9.   Maximize P(x,y) = 140x + 150y

subject to

$$x \geq 0, y \geq 0$$
$$8x + 5y \leq 2210$$
$$3x + 2y \leq \ \ 860$$
$$x \geq 50, y \geq 50$$

10.   Maximize P(x,y) = 180x + 120y

subject to

$$x \geq 0, y \geq 0$$
$$4x + 3y \leq 320$$
$$5x + 2y \leq 330$$

11. Maximize P(x,y) = 190x + 110y

subject to

$$x \geq 0, y \geq 0$$
$$5x + 2y \leq 330$$
$$3.25x + 2y \leq 225$$
$$4x + 3y \leq 320$$

7.4 OTHER LINEAR PROGRAM SOLUTION METHODS

The corner point method is computationally effective for solving 2-variable linear programs. With suitable refinements it can be extended to linear programs in more than two variables, but it very rapidly loses computational effectiveness as the number of variables increases. At this point the simplex method, devised by George Dantzig in the late 1940s, is the most computationally effective general method for solving the widest variety of linear programs.

7.5 THE SCOPE OF LINEAR PROGRAMMING APPLICATIONS

Linear programming has turned out to have a wide spectrum of applications. To obtain some sense of this spectrum we look at six case studies. The case studies are all realistic, but are presented in miniature for the sake of manageability. Actual real-life situations that emerge have the same structure and tone, but are more complex in that more factors are generally considered and more variables are required. Some situations different from the following six are explored in the food for thought exercises.

We view all of these situations through the eyes of others in much the same way that we see events through the eyes of a reporter or observer by reading his account of them in a newspaper, journal or book. Just as the reporter has selected what he believes are important features surrounding the events and has omitted those he considers unessential, we too are looking at features considered crucial to the situations we examine as seen by someone who has made such a selection. This selection reflects assumptions that have been made. To maintain a proper perspective on this it is important to keep in mind that other analysts, as other reporters, might see things in a different light and accordingly make other assumptions.

CASE 1 PRODUCTION PLANNING

The Austin Company's problem of determining the number of DT-1 and DT-2 digital tape players to be made per week so as to maximize profit, considered in Section 7.1, is a production scheduling problem which we expressed in linear program terms under the assumptions introduced. The background that led to the Austin Company's linear programs is illustrative of situations with the following general features: A firm makes a number of products or models of a product and utilizes a number of resources in their manufacture, such as raw materials, labor, capital, different machines, storage facilities. It is assumed that for each product made a fixed amount of each resource is required to make a unit of that product. Within the production time frame a fixed amount of each resource is available and cannot be exceeded. It is also assumed that for a range of possible output levels there is a fixed profit per unit of each product which does not depend on the number of units produced. Under these conditions the problem of determining output levels of the products produced so as to maximize total profit can be formulated in terms of a linear program.

CASE 2 DIET PROBLEMS

A sack of animal feed is to be put together from linseed oil meal and hay. It is required that each sack of feed contain at least 2 pounds of protein, 3 pounds of fat, and 8 pounds of carbohydrate. It is estimated that each unit (a unit is 30 pounds) of linseed oil meal contains 1 pound of protein, 1 pound of fat, 2 pounds of carbohydrate, and that each unit of hay contains 1/2 pound of protein, 1 pound of fat, and 4 pounds of carbohydrate. Linseed oil meal costs $1.50 per unit and hay costs $1.10 per unit.

The problem is to determine how many units of linseed oil meal and hay should be used to make up a sack of animal feed that satisfies the nutritional requirements at minimal cost.

To translate this problem and the assumptions which underlie it into a linear program, let x and y denote the number of units of linseed oil meal and hay, respectively, to be used in making up a sack of animal feed. The introduction of variables to represent the quantities we are seeking to determine is a humble but essential step in translating the background presented into mathematical form. These variables provide us with the crucial bridge linking the mathematical model with the background; if they are incorrectly or equivocally defined, the bridge they establish is in danger of collapsing.

The basic data are summarized in Table 7.4 .

Table 7.4

	No. units used	Cost per unit	Protein per unit (lbs)	Fat per unit (lbs)	Carbo-hydrate per unit (lbs)
Linseed oil meal	x	$1.50	1	1	2
Hay	y	$1.10	1/2	1	4

Since the total cost equals the cost of linseed oil meal, 1.50x, plus the cost of hay, 1.10y, the cost function to be minimized is

$$C(x,y) = 1.50x + 1.10y.$$

The number of pounds of protein in the mixture equals the amount contributed by the linseed oil meal, x pounds, plus the amount contributed by the hay, (1/2)y pounds. The sack of cattle feed must contain at least 2 pounds of protein, which translates to

$$x + (1/2)y \geq 2,$$

or equivalently

$$2x + y \geq 4.$$

Similarly, the fat and carbohydrate requirements are expressed by the constraints

$$x + y \geq 3$$

and

$$2x + 4y \geq 8,$$

or equivalently

$$x + 2y \geq 4,$$

We thus obtain the linear program model

$$\text{Minimize } C(x,y) = 1.50x + 1.10y$$

subject to

$$x \geq 0, y \geq 0$$
$$2x + y \geq 4$$
$$x + y \geq 3$$
$$x + 2y \geq 4,$$

which has solution (1,2) and minimum value 2.7 (see Section 7.3, Exercise 5, p. 143)

To implement this valid conclusion of the model we would use 1 unit of linseed oil meal (30 pounds) and 2 units of hay (60 pounds). Based on the assumed costs of the ingredients, the anticipated cost of a sack of animal feed is $2.70.

The problem considered illustrates "diet problems" with the following general features: A diet, or food substance, is to be put together from a number of available foods. It is required that the diet be balanced in the sense that it must contain minimal amounts of stated nutrients—proteins, fats, carbohydrates, minerals, vitamins, etcetera. It is assumed that each food unit contains a known fixed amount of each nutritional unit and that the unit prices of the food items are known and fixed within the time period considered. The problem is to determine the minimal cost diet which satisfies the prescribed nutritional requirements.

CASE 3 ENVIRONMENTAL PROTECTION

The Saxon Company must produce at least 250 thousand tons of paper annually. From the current operating system 10 pounds of chemical residue is deposited into a neighboring water system for each ton of paper produced. The resulting pollution has become a problem of serious concern, and to remain eligible for state tax benefits the Saxon Company must restrict the chemical residue emitted into the state's water system to not exceed 200 tons per year. Two filtration systems, Delta and Beta, have emerged for consideration. It is estimated that the installation of the Delta system would reduce emissions to 2 pounds for each ton of paper produced, and installation of the Beta system would reduce emissions to 1 pound for each ton of paper produced. Capital and operating costs for the Delta and Beta systems have been estimated at $8 and $12, respectively, per ton of paper produced.

The problem is to determine how many tons of paper should be produced subject to the Delta system and how many should be produced subject to the Beta system so that the emissions standard is met at minimal cost.

Let x and y denote the number of tons of paper to be produced annually subject to the Delta and Beta systems, respectively. The cost function to be minimized is

$$C(x,y) = 8x + 12y.$$

The condition that the Saxon Company must produce at least 250 thousand tons of paper annually is expressed by

$$x + y \geq 250,000.$$

The total amount of chemical residue produced annually is the number of pounds produced through use of the Delta system, 2x pounds, plus the amount produced through use of the Beta system, y pounds. Since this cannot exceed 200 tons, we have

$$2x + y \leq 400,000.$$

We thus emerge with the linear program model.

$$\text{Minimize } C(x,y) = 8x + 12y$$

subject to

$$x \geq 0, \, y \geq 0$$
$$x + y \geq 250,000$$
$$2x + y \leq 400,000,$$

which has solution (150000,100000) and minimum value 2,400,000 (see Section 7.3, Exercise 7, p. 144).

To implement this result the Saxon Company would have to produce 150,000 tons of paper annually subject to the Delta system and 100,000 tons of paper annually subject to the Beta system. The anticipated cost would be $2.4 million.

CASE 4 BANK PORTFOLIO MANAGEMENT

The Charles National Bank has assets in the form of loans and negotiable securities which, it is assumed, bring returns of 10 and 8 percent, respectively, in a certain time period. The bank has a total of $60 million to allocate between loans and securities. To meet unanticipated deposit withdrawals the bank maintains a securities balance greater than or equal to 25 percent of total assets. Lending is the bank's most important activity and to satisfy its clients it requires that at least $15 million be available for loans.

The bank wishes to determine, under these conditions, how funds should be allocated to maximize total investment income.

Let x and y denote the amount, in millions of dollars, to be allocated for loans and securities, respectively. The income function to be maximized is

$$I(x,y) = 0.10x + 0.08y.$$

The following constraints emerge:

$x + y \leq 60$: $60 million is available for investment in loans and securities.

$y \geq (1/4)(x + y)$, or equivalently, $-x + 3y \geq 0$: A securities balance greater than or equal to 25% of total assets must be maintained. Note, total assets is defined as the sum of the amounts invested in loans and securities, which is $x + y$.

$x \geq 15$:   At least $15 million must be available for loans.

We thus obtain the linear program model

$$\text{Maximize } I(x,y) = 0.10x + 0.08y$$

subject to

$$x \geq 0, \; y \geq 0$$
$$x + \; y \leq 60$$
$$-x + 3y \geq \; 0$$
$$x \geq 15,$$

which has solution (45,15) and maximum value 5.7 (see Section 7.3, Exercise 6, p. 143).

To implement this result the Charles Bank would have to allocate $45 million to loans and $15 million to securities. The anticipated interest on investment is $5.7 million.

OF INTEREST

A. Broaddus, "Linear programming: A New Approach to Bank Portfolio Management," *Federal Reserve Bank of Richmond: Monthly Review*, vol. 58, No. 11 (Nov. 1972), pp. 3–11. This article provides an introductory nontechnical discussion of linear programming for bank portfolio management.

K.J. Cohen and F.S. Hammer, "Linear Programming and Optimal Bank Asset Management Decisions," *Journal of Finance*, vol. 22 (May 1967), pp. 147–165. This paper describes a linear program model that had been used for several years by Bankers Trust Company in New York to assist in reaching portfolio decisions.

CASE 5 TRANSPORTATION

Heavy duty transformers made by the Thomas Company are to be sent from their plants in Dobsville and Watertown to distribution centers in New York, Chicago and Detroit. There are 100 transformers in Dobsville and 150 transformers in Watertown. The distribution centers in New York, Chicago and Detroit are to receive 75, 90 and 85 transformers, respectively. It costs $50, $52 and $54 to ship a transformer from Dobsville to New York, Chicago and Detroit, respectively. It costs $51, $48 and $50 to ship a transformer from Watertown to New York, Chicago and Detroit, respectively.

The problem is to determine how many transformers should be sent from each plant to each distribution center so that total cost is minimized.

This situation has the special feature that the total number of transformers to be sent from the plants (250) is equal to total number to be received by the destinations. This allows us to analyze the problem in terms of two variables as opposed to six variables (one linking each source to each destination) which would be needed if this equilibrium condition were not satisfied.

Let x and y denote the number of transformers to be sent from Dobsville to New York and Chicago, respectively. Send what remains at Dobsville, $100 - x - y$ transformers, to Detroit. From Watertown we send to New York, Chicago and Detroit the difference between what they should receive and what they have been sent from Dobsville. Thus, from Watertown we send $75 - x$ transformers to New York, $90 - y$ transformers to Chicago, and $85 - (100 - x - y) = x + y - 15$ transformers to Detroit. In summary, we have the shipping schedule shown in Figure 7.15.

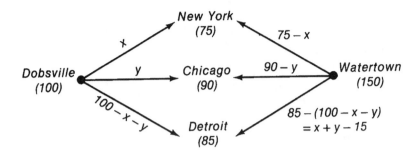

Figure 7.15

The cost of shipping the transformers from Dobsville to New York is 50x dollars—the cost of shipping one transformer, $50, times the number being sent, x. Similarly, the cost of shipping the transformers from Dobsville to Chicago and Detroit is 52y and 54(100 − x − y) dollars, respectively. The cost of shipping the transformers from Watertown to New York, Chicago and Detroit is 51(75 − x), 48(90 − y) and 50(x + y − 15) dollars, respectively.

The total cost function C(x,y), obtained by adding up the costs of shipping the transformers from the sources to the destinations is

$$C(x,y) = -5x + 12{,}795.$$

Since we have taken into account the number of transformers at each source and the number to be received by each destination, the only condition that remains to be stated is that the direction of the flow is from the sources to the destinations; no backflow. Doing so yields the linear program model

$$\text{Minimize } C(x,y) = -5x + 12{,}795$$

subject to

$$x \geq 0, \ y \geq 0$$
$$100 - x - y \geq 0$$
$$75 - x \geq 0$$
$$90 \quad - \quad y \geq 0$$
$$x + y - 15 \geq 0,$$

or equivalently,

$$\text{Minimize } C(x,y) = -5x + 12{,}795$$

subject to

$$x \geq 0, \ y \geq 0$$
$$x + y \leq 100$$
$$x + y \geq \quad 15$$
$$x \leq 75, \ y \leq 90,$$

which has solutions (75,0) and (75,25) and minimum value 12,420. Moreover, all points of the form (75,y), where y is an integer value between 0 and 25, are also solutions of this linear program which could be implemented as a shipping schedule (see Section 7.3, Exercise 8, p. 144).

Implementation of the solution (75,0), for example, requires that 75, 0 and 25 transformers be sent from Dobsville to New York, Chicago, and Detroit, respectively, and that 0, 90 and 60 transformers be sent from Watertown to New York, Chicago, and Detroit, respectively. The anticipated total cost is $12,420.

More generally, a transportation problem has the following features. Given amounts of a commodity are available at a number of sources of supply, such as warehouses. Specified amounts are required by various destinations, such as retail outlets. The total amount required by the destinations may or may not be equal to the total amount available at the sources. Estimates (assumptions) are available on the cost of sending one unit of the commodity from each source to each destination. The problem is to determine the least cost shipping schedule. When the problem involves very few sources and destinations, as in Case 5, it can be solved by inspection. When the number of sources and destinations is large, inspection will not do, and linear programming provides us with a systematic approach to such problems.

CASE 6 AN ASSIGNMENT PROBLEM

The Rasa Publishing Company has two positions to fill, editor of the mathematics list (job 1) and editor of the social science list (job 2), and is considering three candidates, Albert Roberts (candidate 1), Rita O'Brien (candidate 2), and Martin Thorp (candidate 3). After considering resumes, letters of recommendation, and conducting interviews, the editorial board of the company assigned a numerical rating to each person's qualifications for each position as stated in Table 7.5. These ratings serve as a quantitative measure of each candidate's potential for each position as seen by the editorial board. The editorial board wishes to assign candidates to positions in such a way that total potential is maximized.

Table 7.5

Candidate	Position	
	Math, Editor (job1)	Soc. Sci. Editor (job 2)
Roberts (candidate 1)	8	8
O'Brien (candidate 2)	7	9
Thorp (candidate 3)	9	8

To relate the candidates to the jobs we introduce X_{ij} to relate candidate i to job j. X_{ij} can assume one of two values, 0 if candidate i is not assigned job j, and 1 if candidate i is assigned job j.

In summary, we emerge with Table 7.6.

Table 7.6

Candidate	Position	
	Math, Editor (job1)	Soc. Sci. Editor (job2)
Roberts (candidate 1)	X_{11}	X_{12}
O'Brien (candidate 2)	X_{21}	X_{22}
Thorp (candidate 3)	X_{31}	X_{32}

The function

$$P = 8X_{11} + 8X_{12} + 7X_{21} + 9X_{22} + 9X_{31} + 8X_{32},$$

obtained by multiplying the variable that relates a candidate to a job by the candidate's potential for the job and adding, is the potential function to be maximized subject to two conditions:

1. Each candidate is assigned to at most one job.

The variables X_{11} and X_{12} (row 1 of Table 7.6) relate candidate 1 to the available jobs 1 and 2. The constraint

$$X_{11} + X_{12} \leq 1$$

expresses the requirement that candidate 1 be assigned to at most one job since it makes it impossible for candidate 1 to be assigned job 1 ($X_{11} = 1$) and job 2 ($X_{12} = 1$). This constraint comes from row 1 of Table 7.5, and in general the condition that each candidate be assigned at most one job is expressed by requiring that the sum of the variables in each row of Table 7.6 be less than or equal to one. Rows 2 and 3 yield the same condition for candidates 2 and 3:

$$X_{21} + X_{22} \leq 1$$
$$X_{31} + X_{32} \leq 1$$

2. Each job is filled by at most one person.

The variables X_{11}, X_{21}, and X_{31} in column 1 of Table 7.6 relate candidates 1, 2, and 3 to job 1. The constraint

$$X_{11} + X_{21} + X_{31} \leq 1,$$

obtained by requiring that the sum of the variables in the first column 1 of Table 7.6 be less than or equal to one, expresses the requirement that job 1 be filled by at most one candidate since it makes it impossible for any two or all three candidates to be assigned to job 1 ($X_{11} = 1$, $X_{21} = 1$, $X_{31} = 1$). Column 2 yields the same condition for job 2:

$$X_{12} + X_{22} + X_{32} \leq 1$$

We thus obtain the following linear program: find non-negative integers (zeros and ones) that

Maximize $P = 8X_{11} + 8X_{12} + 7X_{21} + 9X_{22} + 9X_{31} + 8X_{32}$

subject to

$$X_{11} + X_{12} \qquad\qquad\qquad\qquad\qquad \leq 1$$
$$X_{21} + X_{22} \qquad\qquad\qquad \leq 1$$
$$X_{31} + X_{32} \quad\; \leq 1$$
$$X_{11} \qquad\quad + X_{21} \qquad\quad + X_{31} \qquad \leq 1$$
$$X_{12} \qquad\quad + X_{22} \qquad\quad + X_{32} \quad \leq 1.$$

By inspection we can see from Table 7.6 that the potential function P is maximized when candidate 3 (Thorp) is assigned to job 1 (Math Editor) and candidate 2 (O'Brien) is assigned to job 2 (Social Science Editor). That is, $X_{11} = 0$, $X_{12} = 0$, $X_{21} = 0$, $X_{22} = 1$, $X_{31} = 1$, $X_{32} = 0$, maximizes potential. When the number of candidates and jobs is large such problems cannot be handled by inspection, but can be handled by linear programming methods.

Assignment problems may involve the most efficient assignment of people to jobs, machines to tasks, project leaders to projects, police cars to city sectors, departments to store locations, sales people to territories, and so on. The objective might involve maximizing effectiveness in some sense or minimizing cost or travel time.

Food for Thought

The situations presented in the following exercises reflect assumptions made by some individual or group. In the first fifteen cases set up linear program models for the problems that arise, solve them, and describe how the solutions obtained would be implemented. What concerns would you want to have satisfactorily addressed before implementing each of the solutions obtained? Do the situations described in Cases 11 and 15 lead to any special difficulties or problems? Set up linear program models for the situations described in 16, 17, 18.

1.   As noted in Exercise 2 of Section 7.3 (p. 140), the Onute Corporation plans to introduce two high resolution TV models, T20 and T24 into the markets. The Company's operation research department was asked to determine the number of units of each model that should be made to maximize profit.

The OR department focused on the manufacturing process which they divided into two phases, construction and finishing. The data collected and their analysis led them to make the following assumptions: In the construction phase each T20 unit requires 4 hours of labor and each T24 unit requires 3 hours of labor. At most 320 hours of construction time are available per day. In the finishing phase each T20 unit requires 5 hours of labor and each T24 unit requires 2 hours of labor. At most 330 hours of finishing time are available per day. The anticipate profit for each unit is $180 and the anticipated profit for each T24 unit is $120.

The problem is to set up a linear program model, M1, reflecting these conditions and underlying assumptions and determine the optimal production schedule for maximizing profit.

2.   As noted in Exercise 2 of Section 7.3 (p. 140), the Onute Corporation also hired the Alexis Company to study its production scheduling problem. The Alexis Company divided the production process into three phases, construction, finishing, and management. They came up with the same assumptions and conditions as the OR department of the Onute Corporation concerning construction and finishing. They were further led to assign 3.25 hours of management time to the production of each unit of T20 and 2 hours of management time to the production of each unit of T24. At most 225 hours of management time were determined as being available per day for the production of T20 and T24. The anticipated profit per unit was determined as $190 for T20 and $110 for T24.

a. The problem is to set up a linear program model, M2, reflecting these conditions and underlying assumptions and determine the optimal production schedule for maximizing profit.

b. Returning to the setting of Exercise 2 of Section 7.3 (p. 140), with further specific details about models M1 and M2 before us, are there any other questions that you would put to the two operations research groups who constructed M1 and M2 before making a decision about whether to implement M1 or M2?

3. A fruit juice is to be made from orange juice concentrate and apricot juice concentrate. Particular attention is being paid to the vitamin A, C, and D content of the fruit juice. Each container of fruit juice is to contain at least 120 units of vitamin A, 150 units of vitamin C, and 55 units of vitamin D. One ounce of orange juice concentrate contains 2 units of vitamin A, 3 units of vitamin C, and 1 unit of vitamin D. One ounce of apricot juice concentrate contains 3 units of vitamin A, 2 units of vitamin C, and 1 unit of vitamin D. Orange juice concentrate costs 3 cents per ounce and apricot juice concentrate costs 2 cents per ounce.

The problem is to determine how many ounces of each concentrate should be used to make a least-cost container of juice that satisfies the vitamin requirements.

4. To advertise its new best seller, the Rasa Publishing Company is planning to buy morning and afternoon time on radio station WQRX. Morning time costs $1000 per minute and afternoon time costs $800 per minute. It is estimated that morning commercials reach 0.9 million listeners and afternoon commercials reach 0.6 million listeners. At most 16 minutes of morning time is available in the month in which the advertising campaign is to run. The advertising department of the Rasa Company feels that at least 8 minutes of morning time and at least 6 minutes of afternoon time should be purchased. The advertising budget for this campaign is $24,000.

How much morning and afternoon time should be purchased so as to maximize the total number of listeners reached in the month in which the advertising campaign is to run?

5. At Ecap university discussion has centered on determining the number of openings, called slots, to be made available in the forthcoming year at the associate and full-professor ranks. Each person promoted to associate professor is to receive a merit increment of $5000, and each person promoted to full-professor is to receive a merit increment of $10,000. At most $150,000 is available for merit increments. A long-standing guideline is that the number of full-professor slots is not to exceed one fourth the number of associate professor slots.

The university senate has recommended that at least 3 slots at the full-professor rank be established and that not more than 22 slots at the associate professor rank be established.

Of interest to the faculty council is the question of how many slots should be established at each rank so as to maximize the total number of promotions. Administration has raised the question of how many slots at each rank should be established so as to minimize the total cost of increments.

6. The Andrius Bank has assets in the form of loans and negotiable securities. For a certain time period it is assumed that loans and securities bring returns of 9 and 6 percent, respectively. The bank has a total of $25 million, provided by demand deposit accounts and time deposit accounts, to allocate between loans and securities. To meet unanticipated deposit withdrawals the bank always maintains a securities balance equal to or greater than 20 percent of total assets. Since lending is the banks most important activity, it imposes certain restrictions on its loan balance to satisfy its principal clients. Specifically, it requires that at least $8 million be available for loans.

   Under the given conditions, how should the bank allocate funds between loans and securities so that total investment income is maximized?

7. Soybeans are to be shipped from New York and New Orleans to Dakar, Marseille, and Odessa. 8000 tons are in New York and 11,000 tons are in New Orleans. 6000 tons are to be sent to Dakar, 7000 tons to Marseille, and 6000 tons to Odessa. The cost (in dollars) of shipping one ton of soybeans from each distribution point to each destination is given in Table 7.7.

Table 7.7

	Dakar	Marseille	Odessa
New York	10	11	12
New Orleans	10.5	11.2	12.3

   The problem is to determine how many tons should be shipped from each distribution center to each destination so that the total shipping cost is minimized.

8. The Brooks and Darius mines of Lexington Mines, Inc., produce high-grade and medium-grade silver ore. The Brooks mine yields 1 ton of high-grade ore and 4 tons of medium-grade ore per hour. The Darius mine yields 2 tons of high-grade ore and 3 tons of medium-grade ore per hour. To meet its commitments the company needs at least 40 tons of high-grade and 100 tons of medium-grade ore per day. It costs $500 per hour to operate the Brooks mine and $700 per hour to operate the Darius mine.

Lexington Mines, Inc., would like to determine how many hours per day each mine should be operated if their ore requirements are to be met at minimum cost.

9. The Stillwell Company produces lubricating oil for machine tools (BV3 oil) and steel mills (BV7 oil). The production of a unit of BV3 oil requires 1 hour of refining and 0.5 hours of blending; the production of a unit of BV7 oil requires 1.2 hours of refining and 1 hour of blending. The refinery can be kept in operation at most 12 hours per day, and the blending plant can be kept in operation at most 8 hours per day. BV3 oil brings a profit of $15 per unit and BV7 oil brings a profit of $20 per unit.

   The problem is to determine the number of units of each type of oil that should be produced daily to maximize profit.

10. The Clinton Company produces a portable X-ray unit that it sells to two types of outlets, hospitals and medical supply houses. The profit margin varies between the two types of outlets owing to differences in order sizes, selling costs, and credit policies. It is estimated that the profit per unit is $100 and $120, respectively, for the hospitals and medical supply houses. Sales promotion is carried out by personal sales force calls and media advertising. The company has eight salespersons on its marketing staff, representing 12,000 hours of available customer contact time during the next year. $30,000 has been allocated for media advertising during the next year. An examination of past data indicates that a unit sale to a hospital requires about a 1/2-hour sales call and $2 worth advertising; a unit sale to a medical supply house requires a 1-hour sales call and $1 worth of advertising. The company would like to achieve sales of at least 5000 units in each customer segment.

   The problem is to determine the sales volume (in product units) it should seek to develop in each customer segment in order to maximize total profit.

11. The Hoffman Clothing Manufacturers, Inc., has available 120 square yards of cotton and 100 square yards of wool for the manufacture of coats and dresses. Two square yards of cotton and 4 square yards of wool are used in making a coat while 4 square yards of cotton are used in making a dress. Cotton costs $5 per

square yard and wool costs $20 per square yard. Four hours of labor are needed to make a coat and 2 hours of labor are needed to make a dress. The cost of labor is $25 per hour. At most 110 hours of labor are available for the manufacture of the coats and dresses.

If a coat sells for $300 and a dress sells for $140, how many of each should be made if net income is to be maximized?

12. The Petrovski Steel Company produces 2 million tons of steel annually. In the current operation of the blast and open-hearth furnaces 50 pounds of particulate matter is emitted into the atmosphere for every ton of steel produced. The resulting air pollution has become a problem of serious concern and efforts are being directed at curbing the emissions. On the basis of studies that have been conducted, it is estimated that installation of the F14 filter system would reduce emissions to 20 pounds of particulate matter per ton of steel produced, and installation of the F24 filter system would reduce emissions to 18 pounds of particulate matter per ton of steel produced. Capital and operating costs for the F14 and F24 filter system are estimated at $1.2 and $1.8 respectively, per ton of steel produced. It is desired that particulate emissions be reduced by 62,400,000 pounds or better per annum. At the same time cost is an important factor if the company is to remain competitive.

    The problem is to determine how many tons of steel should be produced annually subject to the F14 system and how many tons should be produced subject to the F24 system so that the desired reduction in particulate emissions is achieved at minimal total cost.

13. Jones Farms, Inc., has 50 acres available to plant peas and carrots. The estimated average selling price of these vegetables is 20¢ a pound for peas and 11¢ a pound for carrots. The expected average yield per acre is 1000 pounds of peas and 1500 pounds of carrots. The estimated labor needed for sowing, cultivating, and so on, per acre is 8 labor-days for peas and 6 labor-days for carrots. At most 360 labor-days of labor are available at a cost of $20 per labor-day. Fertilizer costs 8¢ per pound, and 100 pounds per acre are needed for peas and 80 pounds per acre are needed for carrots.

    How many acres of peas and carrots should be planted so that income (income = amount derived from sales minus cost) is maximized?

14. The Jay Toy Store plans to invest up to $2200 in buying and stocking two popular children's toys. The first toy costs $4 per unit and occupies 5 cubic feet of storage space; the second toy costs $6 per unit and occupies 3 cubic feet of storage space. The

store has 1400 cubic feet of storage space available. The owner expects to make a profit of $1.50 on each unit of the first toy he buys and stocks and a profit of $2.00 on each unit of the second toy.

How many units of each should be bought and stocked so that profit is maximized?

15. The Inter-City Bus Company plans to invest up to $3 million on new equipment. Two bus models, B4 and B9, are being considered for purchase. Bus B4 is expected to average 16 hours a day at 55 miles an hour with an average of 45 passengers. Bus B4 costs $120,000. Bus B9, a double-decker, is expected to average 18 hours a day at 50 miles an hour with an average of 60 passengers. Bus B9 costs $180,000. The Company wishes to purchase at least 30 new buses.

How many vehicles of each model should be purchased so that capacity in passenger-miles per day is maximized?

16. A legal advisory group is to make recommendations on two positions, State Supreme Court Judge and Civil Court Judge, and is considering three candidates, M. Jones, R. Johnson, and A. Marks. Table 7.8 describes potential ratings that have been assigned by the advisory group as a quantitative measure of each person's qualifications for each position.

Table 7.8

CANDIDATE	JOB	
	Supreme Court Judge	Civil Court Judge
Jones	9	8
Johnson	8	9
Marks	10	8

The advisory group wishes to make its recommendations on how these positions should be filled on the basis of maximization of potential.

17. The Birute Investment Company has $50,000 available for investment. Four stocks, S_1 (oil), S_2 (computers), S_3 (airlines), S_4 (steel), are being considered. Table 7.9 specifies the current price per share of each stock and the expected net profit per share of stock. At most 200 shares of S_1 stock, 300 shares of S_2 stock, 400 shares of S_3 stock, and 300 shares of S_4 stock are to be purchased.

The problem is to determine the number of shares of each stock that should be purchased so as to maximize the total return.

Table 7.9

Stock	Price	Profit
S_1 (oil)	$75	$6.00
S_2 (computers)	$80	$5.00
S_3 (airlines)	$50	$4.00
S_4 (steel)	$60	$4.50

18. Two patients, Ann Levy and Walter Rudd, at General Hospital require blood transfusions. Blood supply data are given in Table 7.10 and data on the requirements of the patients are given in Table 7.11.

The problem is to determine a distribution mechanism which gives each person the required amount of blood in such a way that the replacement cost is minimized. Blood type AB is a universal recipient and type O is a universal donor.

Table 7.10

Blood Type	Supply (pints)	Replacement (per pint)
A	2	$30
B	3	$25
O	2	$20

Table 7.11

Patient	Blood Type	Amount Required (pints)
Levy	A	3
Rudd	AB	2

PROBABILITY MODELS

8.1 PREFACE TO PROBABILITY

By a **random process,** or **experiment,** we mean a process which gives rise to an outcome, but which outcome cannot be predicted with any certainty in advance. The processes of tossing a coin, tossing a die, and choosing a sample from a shipment of items are simple examples of random processes. Although the outcome of any repetition or occurrence of a random process cannot be predicted with any certainty in advance, if we focus on an event of interest over a long series of repetitions of the random process we find stability in the relative frequency with which the event occurs. This stability can serve as a basis for understanding and predicting the behavior of the random process.

To develop this point of view we introduce the following definition.

Let A denote an event connected with a random process. The **relative frequency of A,** denoted by R(A), is the ratio

$$R(A) = \frac{\text{number of times A occurs.}}{\text{number of times the process is repeated}}$$

Thus, if a coin is tossed 5 times and head shows on 2 of the 5 tosses, the relative frequency of heads for this series of 5 tosses is 2/5 or 0.40. We say that head showed 40 percent of the time in the 5 tosses.

Over the short run, that is, for a small number of repetitions of a random process, the behavior of R(A) for a single toss is greatly influenced by a unit increase in its denominator while the numerator remains the same or increases by one. As a result R(A) is unstable and fluctuates considerably. In Figure 8.1 the relative frequency of the event head shows is plotted (vertical axis) against the number of tosses of a certain nickel for the first 15 tosses (horizontal axis). The relative frequency of this event is rather erratic and varies from 0 to 0.533. In Figure 8.2 the relative frequency of head shows is shown for the last 15 of a series of 200 tosses of the nickel. The relative frequency of this event is rather stable at this point, varying from 0.536 to 0.546, since the denominator of the relative frequency ratio is large and the change brought by an additional toss of the nickel is small.

The desire to develop a mathematical structure that would help us understand and predict the long term relative frequency behavior of events connected with random processes played a leading role in the development of probability as this mathematical instrument.

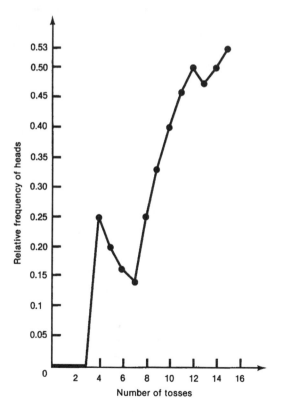

Figure 8.1

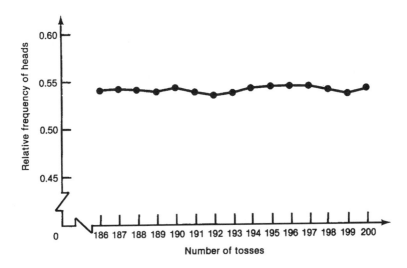

Figure 8.2

SAMPLE SPACE FOR A RANDOM PROCESS

Since our study focuses on the behavior of events, it is desirable to express the nature of the many events that arise in the study of a random process in terms of a set of comparatively few events which is adopted as a foundation. This leads us to the concept of sample space for a random process. To develop this concept consider the process of tossing a die. Some of the events connected with this process are 1 shows, 2 shows, and even number shows, a number between 2 and 5 shows, a number less than 4 shows. To describe events of interest concerning the number which shows when a die is tossed, consider the set of events

S = {1 shows, 2 shows, 3 shows, 4 shows, 5 shows, 6 shows},

which we further abbreviate by writing

S = {1,2,3,4,5,6}.

This set of events has the property that whenever a die is tossed a unique event in S is determined; S is complete in the sense that some event in S occurs when a die is tossed, and unambiguous in the sense that only one event in S occurs when a die is tossed. Other events involving the character of the number showing when the die is tossed can be described in terms of subsets of S. For example, an even number shows is described by $\{2,4,6\}$; a number between 2 and 5 shows is described by $\{3,4\}$ a number greater than 4 but less than 3 shows is described by the humble, but important, empty set $\emptyset$. S = $\{1,2,3,4,5,6\}$ is said to be a sample space for the die tossing process because of its fundamental property that whenever a die is tossed, one and only one of the events in S occurs. S is further said to be a finite sample space because the number of events in it, 6, is a positive integer.

More generally, a collection of events

$$S = \{s_1, s_2, \ldots, s_n\},$$

is said to be a **finite sample space** for a random process if whenever the process is repeated, one and only one of the events in S occurs. The n events $s_1, s_2, \ldots, s_n$ which make up S are call **sample points.** The subsets of S describe events which can be analyzed in terms of S.

In general there are many possible sample spaces which can be given for a random process since all we need for a sample space is a collection of events with the fundamental defining property cited. However, not all sample spaces that may be given for a process are useful in helping us to further understand the process. $S_1 = \{O, E\}$, where O is the event an odd number shows and E is the event an even number shows, is also a sample space for the die tossing process since exactly one of these events occur when a die is tossed; S_1 is not a very useful sample space because there is very little which can be described in terms of it.

Example 1

Specify two sample spaces for the experiment of tossing a pair of dice.

To easily distinguish the dice, we shall assume that one of them is red and the other is green. One sample space, which we shall call S_1, is the set of events listed in Table 8.1. The first number in each ordered pair specifies the number that shows on the red die and the second number specifies the number that shows on the green die. (4,2), for example, is the event that red shows 4 and green shows 2; (2,4) is the event that red shows 2 and green shows 4. S_1 is a sample space because it has the property that whenever the dice are tossed exactly one of the events in S_1 occurs.

Table 8.1

	(1,1)	(1,2)	(1,3)	(1,4)	(1,5)	(1,6)
	(2,1)	(2,2)	(2,3)	(2,4)	(2,5)	(2,6)
$S_1 =$	(3,1)	(3,2)	(3,3)	(3,4)	(3,5)	(3,6)
	(4,1)	(4,2)	(4,3)	(4,4)	(4,5)	(4,6)
	(5,1)	(5,2)	(5,3)	(5,4)	(5,5)	(5,6)
	(6,1)	(6,2)	(6,3)	(6,4)	(6,5)	(6,6)

Another sample space S_2 is

$$S_2 = \{2,3,4,5,6,7,8,9,10,11,12\}$$

where 2 denotes the event that the sum of the numbers showing is 2, 3 denotes the event that the sum of the numbers showing is 3, and so on. S_2 is also a sample space because it too has the property that whenever the dice are tossed exactly one of the events in S_2 occurs.

PROPERTIES OF THE RELATIVE FREQUENCY FUNCTION R(A)

In developing a mathematical structure, called probability model, that would help us understand and predict the behavior of random processes it would make sense to look at properties of R(A) to obtain a sense of direction. Let us recall that for an event A connected with a random process,

$$R(A) = \frac{\text{number of times A occurs}}{\text{number of times the process is repeated}}$$

1. $R(A) \geq 0$. Both numerator and denominator of $R(A)$ are non-negative.

2. $R(A) \leq 1$. The numerator of $R(A)$ cannot exceed its denominator.

3. $R(A) = 0$ if A is an event whose occurrence is not possible (which happens when A is defined by incompatible conditions). Such an event A is identified with $\emptyset$.

4. $R(S) = 1$, where $S = \{s_1, s_2 \ldots, s_n\}$ is a sample space for the random process. Whenever the process is repeated one of the sample points in S occurs, and thus S occurs. Therefore, the numerator and denominator of $R(S)$ are equal.

In connection with the die tossing experiment, consider the event an even number shows. Let us observe that since an even number shows as often as 2, 4, and 6 show, the relative frequency of occurrence of this event is the sum of the relative frequencies of the events 2 shows, 4 shows and 6 shows. If, for example, 2, 4 and 6 are observed to show with relative frequencies 19/100, 27/100, and 14/100, respectively, then the relative frequencies of an even number's occurrence is $19/100 + 27/100 + 14/100 = 60/100 = 60$ percent. Such is the case in general, and we state this observation as follows:

5. R(A) =   sum of the relative frequencies of the sample points that describe A.

6. R(S) = $R(s_1) + R(s_2) + \ldots + R(s_n)$, where $S = \{s_1, s_2, \ldots, s_n\}$ is sample space for the process. This property is obtained by applying property 5 to S.

7. $R(s_1) + R(s_2) + \ldots + R(s_n) = 1$; the sum of the relative frequencies of all sample points is 1. This follows from properties 4 and 6.

Food for Thought

1. Toss a coin 200 times, record the occurrence of the event tail shows, and determine the relative frequency of tail shows throughout the sequence of 200 tosses. Draw graphs to show the fluctuation in relative frequency for the first 15 tosses, tosses 86 through 100, and tosses 186 through 200.

2. Consider the process of tossing a die.

   a. Let L denote the event that a number less than 3 shows, 4 denote the event that 4 shows, and B denote the event that a number greater than 4 shows. Is $S = \{L,4,B\}$ a sample space for this process? How so?

   b. Let E denote the event that an even number shows, and A denote the event that a number greater than 3 shows. Is $S_1 = \{E,A\}$ a sample space for this process? Explain.

   c. Let A denote the event that a number less than 3 shows, B denote the event that a number between 3 and 5 shows, and C denote the event that a number greater than 4 shows. Is $S_2 = \{A,B,C\}$ a sample space for the process? How so?

3. Consider the process of tossing a pair of dice. Let A denote the event that the sum of the numbers showing is less than 5, B denote the event that the sum of the numbers showing equals 5, and C denote the event that the sum of the numbers showing is greater than 5. Is $S = \{A,B,C\}$ a sample space for this process. How so?

4. Set up two sample spaces for the process of tossing a coin twice in succession.

5. Set up three sample spaces for the process of dealing a card from a standard deck of 52 cards.

6. The Twolow Company makes light bulbs. Two plants, P1 and P2, carry out the production process. The daily output is 8000 bulbs with P1 producing 5000 bulbs of which 1 percent are defective, and P2 producing 3000 bulbs of which 0.5 percent are defective: A bulb is selected from the day's output. Set up three sample spaces for the selection process.

7. A college residence provides housing for 4 students, 3 of which are majoring in history. A sample of 2 students is chosen from the residence. Set up two sample spaces for the selection process.

8.2 STRUCTURE OF A PROBABILITY MODEL

Since the concept of probability model is intended to be applicable to the study of the long term relative frequency behavior of events connected with random processes, the manner in which we define this concept is guided by properties of relative frequency.

A **(finite) Probability model** for a random process consists of two components:

i.   A sample space $S = \{s_1, \ldots, s_n\}$.

ii. A function P, called a **probability function**, which assigns to each subset A of S a value, denoted by P(A) and called the probability of A, subject to the conditions listed below. We state these conditions on the sample points and then extend them to subsets of S.

1. $P(s_1) \geq 0, \quad P(s_2) \geq 0, \ldots, P(s_n) \geq 0$

2. $P(s_1) \leq 1, \quad P(s_2) \leq 1, \ldots, P(s_n) \leq 1$

3. $P(s_1) + P(s_2) + \ldots + P(s_n) = 1$

4. If A is a subset of S, P(A) = sum of the probabilities of the sample points describing A; if A = Ø, P(A) = 0. (For example, if A = {s_1, s_5}, P(A) = P(s_1) + P(s_5).)

The structural requirements of a probability model for a random process are somewhat analogous to a community's building code requirements for building a house. A building code does not tell us how to build a house. It tells us that however we build our house, for it to be legitimate in terms of the building code it must satisfy such and such conditions which are spelled out in the code. The concept of probability model does not tell us how to define a sample space and probability function for a random process; it tells us that in building a specific probability model for a random process, which can be done in many ways, we must satisfy the conditions stated in the definition of probability model (the building code in this case) in order for the model to be mathematically legitimate.

To take an example, Janet James was presented with the following structure which was claimed to be a probability model for the process of tossing a die.

$$S = \{O,E\}, \quad P(O) = \frac{1}{3}, \quad P(E) = \frac{2}{3},$$

where O is the event that an odd number shows and E is the event that an even number shows on a throw of the die. Is this structure a probability model?

S is a sample space for the die tossing process since it satisfies the requirement that exactly one of its events occur when the die is tossed. The assignment by P of the numerical values to O and E satisfies the requirements of a probability function since 1/3 and 2/3 are non-negative, less than 1, and sum to 1. This establishes the mathematical legitimacy of S, P as a probability model for the die tossing process; whether or not this model realistically describes the behavior of any die in Janet's possession is another question entirely. We turn to the issue of formulating a specific probability model for a specific random process in the next two sections.

At first sight the concept of probability model seems no different from the properties of relative frequency that were listed, except for the notation used. This is not the case. To be sure, the conditions required of a probability function closely mirror properties of relative frequency, but relative frequency is defined in a very specific way while probabilities can be assigned to events in a wide variety of ways so long as the conditions cited are satisfied. Probability assignments reflect properties of relative frequency, but go beyond them in much the same way that a son may reflect properties of his father, but goes beyond them.

TWO THEOREMS OF PROBABILITY

There are a number of valid consequences which come out of the probability model structure, but the two theorems considered here are of particular interest to us.

Theorem 1. Let $S = \{s_1, s_2, \ldots, s_n\}$ denote a sample space with n sample points. Let us suppose that for one reason or another we are led to the probability function P which assigns the same value x to all of the sample points in S; then this value is $x = 1/n$.

Proof: From the definition of probability model, we have:

$$P(s_1) + P(s_2) + \ldots + P(s_n) = 1$$

Since each of $P(s_1)$, $P(s_2)$, $\ldots$, $P(s_n)$ equals x, we obtain:

$$\underbrace{x + x + \ldots + x}_{n \text{ terms}} = 1$$

$$nx = 1$$

$$x = \frac{1}{n}$$

A probability model in which all sample points are assigned the same probability value is called an **equally likely outcome** model.

Theorem 2. If $S = \{s_1, s_2, \ldots, s_n\}$, $P(s_1) = \ldots = P(s_n) = 1/n$, and A is an event that is described by k sample points, then:

$$P(A) = \frac{k}{n}$$

Proof: For the sake of simplifying our discussion, let us suppose that A is described by the first k sample points $s_1, s_2, \ldots, s_k$. Then we have:

$$P(A) = P(s_1) + P(s_2) + \ldots + P(s_k)$$

$$= \underbrace{\frac{1}{n} + \frac{1}{n} + \ldots + \frac{1}{n}}_{k \text{ terms}} = \frac{k}{n}$$

In the special case of equally likely outcomes probability questions reduce to counting questions. To determine the probability of A in this framework, count the number of sample points that describe A, count the number of sample points, and take their ratio. If S contains 10,000 sample points and A is described by 1000 of them, then $P(A) = 1000/10{,}000 = 1/10$.

8.3 A TALE OF THREE PROBABILITY MODELS

The following story might at first sight seem rather outrageous, but when we consider the rather weird doings which are reported every day in the press and on television, it's not at all outrageous and is in fact perfectly possible.

Rasa Adams's friend Jean was on the eve of her twenty-first birthday for which a big family celebration was planned. Since Jean was interested in the laws of chance, her mother decided to obtain a golden die as a birthday present for her. On learning of this in a conversation with Jean's mother, Rasa concluded that the best present she could get Jean to go with the golden die would be a probability model; after all, what good is a golden die without an accompanying probability model? But what's the best place to shop for a probability model, thought Rasa. She finally decided to try Martin's Models, run by the noted model builder Robert J. Martin. Rasa was not disappointed at Martin's Models. There was a large selection with many attractive models, and Robert Martin was willing to custom design a model to your specifications if you so desired. Rasa finally narrowed her search to three models, R4 (red), B7 (blue) and Y3 (yellow). These models had the following specifications:

R4 (red): $S = \{1,2,3,4,5,6,\}$; $P(1) = \ldots = P(6) = 1/6$

B7 (blue): $S = \{1,2,3,4,5,6,\}$; $P(1) = P(3) = P(5) = 1/9$

$$P(2) = P(4) = P(6) = 2/9$$

Y3 (yellow): $S = \{1,2,3,4,5,6,\}$; $P(1) = P(3) = P(5) = 1/12$

$$P(2) = P(4) = P(6) = 3/12$$

This is as far as Rasa can go in selecting a probability model for Jean's die. To determine which of these models, if any, is a realistic fit to Jean's die, she needs to know more about the nature of this die. Rasa called Jean's mother and asked this question. "I had to pay extra for this," said Jean's mother, "but this is an unusual die in that platinum weights have been inserted inside it so that the even numbers are favored to show over the odd ones by 2 to 1." This led Rasa to purchase the blue model B7 as the most realistic model for Jean's die.

To examine a valid consequence of the blue model, consider the event E that an even number shows. $E = \{2,4,6\}$.

$$P(E) = P(2) + P(4) + P(6) = \frac{2}{9} + \frac{2}{9} + \frac{2}{9} = 0.67$$

The relative frequency interpretation of this result is that if a die whose behavior is described by the blue model is tossed a large number of times, an even number will show approximately 67% of the time.

After Jean's birthday it came to pass that her die was tossed 500 times. Records kept of the outcomes of the tossings reveal that an even number showed on 246 of the 500 tosses, so that R(E) = 0.492. Since the actual relative frequency of an even number showing, 0.492, is very much at variance with the projected value of approximately 0.67, this shows that a valid consequence of Jean's model, interpreted in relative frequency terms, is false. Jean's model is not a realistic one for her die, and her mother paid for an unusual die which was not delivered. The result observed suggests that the red model, based on the assumption that the die in question is well-balanced, of uniform construction, might be a more realistic one for the die that Jean actually had.

Food for Thought

1. Consider the yellow model Y3 for the process of tossing a certain
    die. S = {1,2,3,4,5,6,},

$$P(1) = P(3) = P(5) = \frac{1}{12}; \quad P(2) = P(4) = P(6) = \frac{3}{12}.$$

Let E denote the event that an even number shows.

  a. Find P(E).

  b. State the relative frequency interpretation of the result ob-
      tained in (a).

  c. In tossing the die in question 1000 times, an even number was
      observed to show 665 times. Does this show that the conclu-
      sion obtained in (a) is not valid? How so?

  d. Is the conclusion reached in (a), interpreted in relative fre-
      quency terms, true? How so?

  e. Is the yellow model Y3 realistic for the die in question? How
      so?

2. Consider the model G4 (green) for the process of tossing a certain
    die. S = {1,2,3,4,5,6},

$$P(1) = P(3) = P(5) = \frac{1}{10}, \quad P(4) = \frac{3}{10}, \quad P(2) = P(6) = \frac{2}{10}$$

Let A denote the event that an odd number shows.

  a. Find P(A).

  b. State the relative frequency interpretation of the result ob-
      tained in (a).

c. In tossing the die in question 1000 times an odd number was observed to show 302 times. Does this evidence establish that the conclusion obtained in (a) is valid? How so?

d. Is the conclusion obtained in (a), interpreted in relative frequency terms, true? Explain.

e. Is the green model G4 realistic for the die in question? How so?

8.4 PROBABILITY MODELS FOR RANDOM PROCESSES

Consider the process of dealing a card from a standard deck of 52 cards and let us address the problem of setting up a probability model for this process and determining the probability that a picture card is dealt.

For convenience in referring to the cards, let us set up a translation system so that we can refer to the cards as 1, 2 . . ., 52; in this translation system 1 might denote the ace of spades, 2 the king of spaces, etc. We take as our sample space,

$$S = \{C_1, C_2, \ldots, C_{52}\},$$

where C_1 is the event that card 1 is dealt, C_2 is the event that card 2 is dealt, etc.

Our next task is to define a probability function P on S. This can be done in many ways, and the function P that emerges depends on the assumption we make about how the card will be dealt from the deck.

Suppose we assume what is usually assumed in such situations, but not always made explicit, which is, that the card is dealt from a well-shuffled deck in a unbiased way, at random, as we say. The probability function P which best reflects this assumption assigns the same value, 1/52, to each sample point in S. We thus emerge with the following probability model:

Model 1: $S = \{C_1, C_2, \ldots, C_{52}\}$

$$P(C_1) = \ldots = P(C_{52}) = \frac{1}{52}$$

From Model 1 it follows that the probability that a picture card is dealt is 12/52 or 0.23.

As is well known, some card dealers are less than honest. Suppose we assume, based on past experience, that the dealer intends to "arrange things" so that the card we are dealt is neither a picture card nor an ace, but that any of the other cards may be dealt without bias. For notational convenience suppose that the picture cards and aces are in the cards we numbered 1,2, . . . 16, and that the cards 17, 18, . . . 52 correspond to the remaining cards. This leads to Model 2.

Model 2: $S = \{C_1, \ldots, C_{16}, C_{17}, \ldots, C_{52}\}$,

$$P(C_1) = \ldots = P(C_{16}) = 0, \quad P(C_{17}) = \ldots = \ldots = P(C_{52}) = \frac{1}{36}.$$

From Model 2 it follows that the probability that a picture card is dealt is 0.

SUSAN'S PROBLEM

Susan Reti was interested in determining the probability that an even sum shows for the process of tossing a pair of well-balanced dice of uniform construction (one red and one green). She asked two of her friends, Rachael and Laura, if they would help her set up a probability model to determine the probability of this event. Both were glad to do so.

Rachael's Model. Rachael took as her sample space S_1 the set of events described by all ordered pairs of integers between 1 and 6, inclusive, as shown in Table 8.2. The first number in each ordered pair specifies the number which shows on the red die and the second number specifies the number which shows on the green die.

Table 8.2

	(1,1)	(1,2)	(1,3)	(1,4)	(1,5)	(1,6)
	(2,1)	(2,2)	(2,3)	(2,4)	(2,5)	(2,6)
$S_1 =$	(3,1)	(3,2)	(3,3)	(3,4)	(3,5)	(3,6)
	(4,1)	(4,2)	(4,3)	(4,4)	(4,5)	(4,6)
	(5,1)	(5,2)	(5,3)	(5,4)	(5,5)	(5,6)
	(6,1)	(6,2)	(6,3)	(6,4)	(6,5)	(6,6)

The assumption that the dice are well-balanced is best reflected by the probability function P which assigns the same value, 1/36, to each sample point in S_1.

The sample points with even sums are found in alternate diagonals of Table 8.2, beginning with the first, and are 18 in number. Thus, from Rachael's model if follows that the probability of E, that an even sum shows, is:

$$P(E) = \frac{18}{36} = 0.50$$

The relative frequency interpretation of this conclusion is that if a pair of well-balanced dice are tossed a large number of times, an even sum will show approximately 50% of the time.

Laura's Model. Laura proceeded to analyze Susan's problem in a different way. She took as her sample space

$$S_2 = \{2,3,4,5,6,7,8,9,10,11,12\}$$

where 2 is the event that the sum of the numbers showing is 2, 3 is the event that the sum of the numbers showing is 3, and so on.

The assumption that the dice are well-balanced led Laura to the probability function P which assigns the same value, 1/11, to the eleven sample points in S_2.

$$E = \{2,4,6,8,10,12\}, \text{ and Laura obtained}$$

$$P(E) = \frac{6}{11} = 0.55$$

for the probability that an even sum shows.

The relative frequency interpretation of this result is that if a pair of well-balanced dice are tossed a large number of times, and even sum will show approximately 55% of the time.

Susan was more confused than ever and turned to her cousin Jack for help with her problem.

Jack's Model. Jack took as his sample space S_3 the collection of events shown in Table 8.3.

Table 8.3

{1,1}	{1,2}	{1,3}	{1,4}	{1,5}	{1,6}
	{2,2}	{2,3}	{2,4}	{2,5}	{2,6}
		{3,3}	{3,4}	{3,5}	{3,6}
			{4,4}	{4,5}	{4,6}
				{5,5}	{5,6}
					{6,6}

Here {1,1} is the event that both dice show 1, {1,2} is the event that one die shows 1 and the other shows 2, and so on. There are 21 sample points, Jack observed. Since the dice are assumed to be well balanced, Jack was led to take as his probability function P the one which assigns the same value, 1/21, to each sample point. There are 12 sample points with even sum and this led Jack to conclude that P(E), the probability that an even sum shows, is 12/21 or 0.57.

The relative frequency interpretation of this result is that if a pair of well balanced dice are tossed a large number of times, an even sum will show approximately 57% of the time.

SUSAN'S DILEMMA

Does $P(E) = 0.50$ (Rachael's Model), 0.55 (Laura's Model), or 0.57 (Jack's Model)? If mathematics is such a precise subject, how could this happen? Which result should I accept, Susan pondered? All three conclusions are correct in the sense of being valid consequences of their respective probability models; from the point of view of validity there is no conflict since these conclusions arise from different sources.

As to the question of truth, these conclusions, interpreted in relative frequency terms, cannot all be true for well-balanced dice. To settle the question of truth, well-balanced dice would have to be tossed a large number of times and a record kept of how often an even sum shows. When this is done, it is found that an even sum shows in the neighborhood of 50 percent of the time. Such evidence, obtained by performing the process and observing, establishes the truth of $P(E) = 0.50$ and the falsity of $P(E) = 0.55$ and $P(E) = 0.57$ in terms of the relative frequency interpretation of probability, for well balanced dice. The appearance of a valid conclusion that is false alerts us to the unrealistic nature of Laura's assumption of equally likely outcomes for S_2. Indeed, it is not difficult to pinpoint the difficulty with Laura's assumption. It is unrealistic to assume, for example, that a sum of 2 is as likely to occur as a sum of 7 when well-balanced dice are tossed; a sum of 2 can only occur in one way—when $(1,1)$ shows; a sum of 7 can occur in six ways—when $(1,6)$ $(6,1)$, $(2,5)$, $(5,2)$, $(3,4)$ or $(4,3)$ show.

Food for Thought

1. The appearance of a valid conclusion that is false also alerts us to the unrealistic nature of Jack's assumption of equally likely outcomes for S_3.

   a. Pinpoint the difficulty with Jack's assumption.

   b. How should Jack's probability function be modified to make it realistic for well-balanced dice?

2. How should Laura's probability function be modified to make it realistic for well-balanced dice?

3. The Twolow Company makes light bulbs. Two plants, P1 and P2, carry out the production process. The daily output is 8000 bulbs of which 1 percent are defective, and P2 producing 3000 bulbs of which 0.5% are defective. A bulb is selected from the day's output.

    a. When asked to determine the probability that a defective bulb is chosen, Mark Twolow set up the following probability model for the selection process, based on the assumption that the bulb is selected at random from the day's output. S = {GP1, GP2, DP1, DP2}, where GP1 is the event that a good bulb made by P1 is selected, and so on.

    $$P(GP1) = P(GP2) = P(DP1) = P(DP2) = 1/4$$

    Mark determined the probability that a defective bulb is chosen to be P(DP1) + P(DP2) = 1/2. (1) Is Mark's model satisfactory? How so? (2) Is Mark's conclusion correct? How so?

    b. Mark's brother Bob suggested a simpler model, again based on the assumption that the bulb is chosen at random from the day's output. S = {G,D}, where G is the event that a good bulb is selected and D is the event that a defective bulb is chosen. P(G) = P(D) = 1/2. Bob pointed out that he and Mark had reached the same conclusion, so that each confirms the correctness of the other.

        1. Would you agree or disagree with Bob's comment? Explain.

        2. Is Bob's model satisfactory? Explain.

        3. Is Bob's conclusion correct?

        4. When asked for the probability that a defective bulb made in P1 is selected in terms of his model, Bob gave 1/4 as the answer. Would you agree or disagree? How so?

c. Formulate your own probability model for the light bulb selection process, assuming that the bulb is chosen at random from the day's output.

d. Should Mark's probability function be modified? How so? If modification is called for, how would you carry it out?

e. Should Bob's probability function be modified? Explain. If modification is called for, how would you carry it out?

4. Jason took a well-balanced coin from his pocket and asked his friend Andrew to help him determine the probability of throwing one head and one tail on two successive tosses of the coin. Andrew took $S = \{0,1,2\}$, where 0 is the event that no heads show in the two tosses, 1 is the event that one head shows in the two tosses, etc., as his sample space. He defined a probability function P by $P(0) = P(1) = P(2) = 1/3$, so that the probability of throwing one head and one tail is 1/3. The relative frequency interpretation of this conclusion is that if a well-balanced coin is tossed twice in succession a large number of times, a head and tail will show approximately 33.3% of the time. When Jason's well-balanced coin was tossed twice in succession 500 times (which involves 1000 tosses), a head and tail were observed to show 246 times.

a. Does this mean that Andrew's conclusion is not valid? How so?

b. Is Andrew's conclusion, interpreted in relative frequency terms, true? How so?

c. How is the discrepancy between the predicted relative frequency and the observed relative frequency to be explained?

d. Set up your own probability model for tossing a coin, assumed to be well balanced, twice in succession.

e. Should Andrew's probability function be modified? Explain. If modification is called for, how would you carry it out?

5. Bill Albert succeeded in finding a rare pair of loaded dice. With reference to the set S_1 of 36 outcomes described in Table 8.2, these dice have the property that outcomes with an even sum

[such as (1,1), (1,3), and so on] are twice as likely to occur as outcomes with and odd sum [such as (2,1), (3,2), and so on].

a. Define a probability function on S_1 that best reflects the nature of Bill's dice.

b. Find the probability that (i) an even sum shows; (ii) a sum of 7 shows; (iii) a sum less than 6 shows.

8.5 RETURN TO EQUALLY LIKELY OUTCOME MODELS

As we have seen, in the special case of an equally likely outcome probability model, wherein all sample points are assigned the same probability value, probability questions reduce to counting questions. If k is the number of sample points describing an event A and n is the total number of sample points, then the probability of A is the ratio k/n. The counting process may range from trivial to extraordinarily complex, and to facilitate the task of counting we look at two basic principles.

TOOLS FOR COUNTING

> **Multiplication Principle.** Let us suppose that a task is to be performed and that it can viewed as a sequence of two procedures where the first can be performed in h ways and, after it has been performed, the second can be performed in k ways; then the two procedures can be performed in the stated order in h · k ways.

Since any one of the h ways in which the first procedure can be performed can be coupled with any of the k ways in which the second procedure can be performed, there are h groups of k possibilities, which gives us h · k outcomes.

More generally, the multiplication principle extends to a sequence of any number of procedures which are to be performed in order.

The challenge to applying the multiplication principle is in seeing a situation from the point of view of a sequence of procedures to be performed in order. Sometimes it is obvious that this is the case, but often this view is more deeply hidden.

Example 1

Two books are to be chosen from three, denoted by A, B and C, and arranged next to each other on a bookshelf. (a) How many arrangements are possible? (b) How many arrangements are possible if second place must be filled by book B?

  a. Two positions are to be filled, which we think of as first place and second place. For first place we can choose any of the three available books and, after this has been done, for second place we can choose any of the two remaining books. This yields $3 \cdot 2 = 6$ possible arrangements, namely, AB, AC, BA, BC, CA and CB.

$$\frac{3}{\text{1st place}} \cdot \frac{2}{\text{2nd place}} = 6$$

  b. If a certain task is to be handled in a special way, it is best to turn to it first. Second place can be filled in one way since it must be filled with B. Turning to first place, we have two options, fill it with A or C. This yields $2 \cdot 1 = 2$ arrangements, AB and CB.

$$\frac{2}{\text{1st place}} \cdot \frac{1}{\text{2nd place}} = 2$$

In both cases the result is obvious, but the approach underlying the analysis is instructive.

Example 2

A license plate is to consist of two capital letters followed by three digits. In how many ways can we construct such a license plate?

We can look at this problem from the point of view of five spaces to be filled in order. In first place we may put any of the 26 capital letters, in second place we may put any of the 26 capital letters, in third place we may put any of the 10 digits, as is the case for fourth and fifth places. Thus, there are $26 \cdot 26 \cdot 10 \cdot 10 \cdot 10 = 676,000$ ways in which we can construct such a license plate.

$$\underbrace{26}_{\text{1st place}} \cdot \underbrace{26}_{\text{2nd place}} \cdot \underbrace{10}_{\text{3rd place}} \cdot \underbrace{10}_{\text{4th place}} \cdot \underbrace{10}_{\text{5th place}} = 676,000$$

Example 3

Alice woke up late and rushed to school without having breakfast. Before going to her second class she decided to make a stop in the school cafeteria for a snack consisting of a sandwich and beverage. She found 5 different kinds of sandwiches and 6 different kinds of beverages to choose from. In how many ways can she put together a snack?

We can think of this in terms of two procedures to be performed, choose a sandwich and then choose a beverage, which can be done in 5 · 6 = 30 ways. Equivalently, choose a beverage and then choose a sandwich, which yields the same result.

Addition Principle. Suppose that a task can be performed in h ways and that a second task can be performed in k ways, where the performance of one of these tasks excludes the performance of the other; then one or the other of these tasks (but not both) can be performed in h + k ways.

More generally, this addition principle extends to a setting involving any number of tasks, where the performance of any one of them excludes the performance of any of the others.

Example 4

As the semester moved along Alice found herself more and more frequently getting up late. One day she went for a snack and found, to her consternation, that she had forgotten her wallet and did not have enough loose change to purchase both a sandwich and a beverage. In how many ways can she select a snack if, as previously, there are 5 different kinds of sandwiches and 6 different kinds of beverages?

Because of financial difficulties, Alice has 5 + 6 = 11 options. She may choose one of the 5 sandwiches or one of the 6 beverages.

Food for Thought

1. In how many ways can 2 of 6 books be chosen and arranged next to each other on a shelf?

2. An encyclopedia of science consists of 7 volumes. (a) In how many ways can these volumes be arranged next to each other on a shelf? (b) How many of these arrangements are out of order? (c) In how many of these arrangements will volume 1 occupy first place and volume 2 occupy second place?

3. A traveler is planning to go from New York to Chicago by plane and make the return trip by bus. There are 5 airlines that have flights at the desired time and 3 bus lines that provide Chicago to New York service. In how many ways can the trip be made?

4. Motors are to pass through two inspection stations. At the first station 2 ratings are possible; at the second station 4 ratings are possible. In how many ways can a motor be marked?

5. Air-conditioners are to be assembled in four stages. At the first stage there are 3 assembly lines, at the second stage 4 assembly lines, at the third stage 4 assembly lines, and at the fourth stage 3 assembly lines. In how many ways can an air-conditioner be routed through the assembly process?

6. In how many ways can 5 people line up at a ticket office?

7. How many 4-digit numbers are there? Note, 0 cannot occupy first place.

8. How many numbers are there between 1000 and 5000, including 1000 and 5000?

9. An examination consists of 8 true-false questions. How many possible different answer sheets can be turned in?

10. Andrius has 7 books, 4 in English and 3 in Lithuanian, which he wants to arrange on a bookshelf. How many arrangements are possible if (a) there are no restrictions; (b) a book in English is to occupy first place and a book in Lithuanian is to occupy second place; (c) the books in English are to occupy the first four places; (d) books in the same language are to be kept together?

11. In how many ways can n people line up at a ticket office?

12. Robert Adams went shopping one day to buy a shirt and pair of shoes. He found 8 shirts and 6 shoe styles to choose from. If he did not have enough funds to pay for both a shirt and shoes, in how many ways could he make a purchase?

13. A signal is a 3-digit sequence of 0's and 1's. How many signals are there?

14. In how many ways can four girls and four boys be seated alternately in a row of eight chairs numbered from 1 to 8 if (a) a boy is to occupy the first chair; (b) either a boy or a girl can occupy the first chair?

15. A telephone dial has 10 holes. How many different signals, each consisting of seven impulses in succession, can be formed (a) if no impulse is to be repeated in any given signal; (b) if repetitions are permitted?

16. In how many ways can a baseball team of 9 players be arranged in batting orders (a) if tradition is followed and the pitcher must bat last; (b) if no restriction is imposed; (c) if a certain 4 players must occupy the first 4 positions in some order?

PERMUTATIONS

A **permutation** of a set of objects is an arrangement of these objects in some order in a line. If there are n distinct objects, then any arrangement of r of them in some order in a line is called a **permutation of the n distinct objects taken r at a time**. The number of permutations of n distinct objects taken r at a time is denoted by $P(n,r)$ or $_nP_r$.

Example 5

Returning to Example 1, which involved choosing 2 books from 3, denoted by A, B, and C, and arranging them next to each other on a shelf, there are $P(3,2) = 3 \cdot 2 = 6$ permutations of the 3 books taken 2 at a time. The permutations are AB, AC, BA, BC, CA and CB.

As a computation tool for $P(n,r)$ we have

$$P(n,r) = n(n - 1) \ldots (n - r + 1),$$

which is the product of the first r integers in descending order starting with n. To establish this result, we note that there are r places to be filled and n objects from which to choose. First place can be filled with any of these n objects. After it has been filled with one of these *n* objects, there remain *n* − 1 objects available for filling the second place. Once it has be filled, 2 objects will have been used and n − 2 choices remain for the third place; once first, second, and third places have been filled, 3 objects will have been used, and n − 3 choices remain for the fourth place. More generally, when we come to the rth place, r − 1 objects will have been used to fill the previous r − 1 places, and n − (r − 1) or n − r + 1 choices remain for rth place. From the multiplication principle, the number of ways of filling the r places is n(n − 1) . . . (n − r + 1).

For example, P(7, 3) is the product of the first 3 integers in descending order beginning with 7, P(7,3) = 7 · 6 · 5 = 210. Also, P(10,4) = 10 · 9 · 8 · 7 = 5040, and P(48,2) = 48 · 47 = 2256.

Since products of consecutive integers arise frequently in counting problems, it is useful to have notation to denote such products. The symbol n! (read "n factorial") is used to stand for the product of all integers from 1 to n inclusive. Thus

$$1! = 1 \qquad\qquad 4! = 4 \cdot 3 \cdot 2 \cdot 1 = 24$$

$$2! = 2 \cdot 1 = 2 \qquad\qquad 5! = 5 \cdot 4 \cdot 3 \cdot 2 \cdot 1 = 120$$

$$3! = 3 \cdot 2 \cdot 1 = 6 \qquad\qquad 6! = 6 \cdot 5 \cdot 4 \cdot 3 \cdot 2 \cdot 1 = 720$$

$$n! = n(n - 1)(n - 2) \ldots 1$$

It is convenient to define 0! by

$$0! = 1.$$

This definition, although perhaps strange at first sight, is useful in that certain counting formulas can be more easily stated without a need for considering separate special cases.

When r = n, we have the following special case: The number of permutations of n distinct objects (taken n at a time) in a line is:

$$P(n,n) = n! = n(n - 1) \ldots 1$$

The computation formulas derived for P(n,r) and P(n,n) follow from the multiplication principle. Many, but not all, counting problems exhibit a structure which permit us to apply these results directly. In other cases which exhibit this structure it is preferable to go back to basic principles to work the problems.

COMBINATIONS

Many situations arise in which r distinct objects are to be selected from n without regard to order. A subset or selection of r objects chosen from n distinct objects, without regard to the order in which they were chosen or appear, is called a **combination of n objects taken r at a time.** The number of combinations of n distinct objects taken r at a time is denoted by C(n,r) or $_nC_r$.

Example 6

Returning to Examples 1 and 5 which involved choosing 2 books from 3, denoted by A, B, and C, we saw that there are 6 permutations of the 3 books taken 2 at a time, namely AB, AC, BA, BC, CA and CB.

There are 3 combinations of these 3 books taken 2 at a time, namely, {A,B}, {A,C} and {B,C}. The combination {A,B} is the set consisting of A and B, no order implied. This combination gives rise to two permutations, AB, which means books A and B in the order A followed by B, and BA, which means books A and B in the order B followed by A.

The number of combinations of r objects chosen from n distinct objects is expressed by the following formula:

$$C(n,r) = \frac{P(n,r)}{r!} = \frac{n(n - 1) \ldots (n - r + 1)}{r!}$$

To establish this result, consider the related problem of determining $P(n,r)$ and think of the process of forming a permutation of r objects selected from n as being carried out in two stages. The first stage consists of selecting r of n objects without regard to order, which can be done in $C(n,r)$ ways. The second stage consists of arranging the r objects chosen in some order, which can be done in $r!$ ways. From the multiplication principle, the number of ways of selecting r of n objects with regard to order, $P(n,r)$, equals the number of ways of selecting r of n objects without regard to order, $C(n,r)$, times the number of ways of ordering the r objects chosen, $r!$. That is:

$$P(n,r) = C(n,r) \cdot r!,$$

Solving for $C(n,r)$ by dividing both sides by $r!$ yields the desired result.

Thus, for example, the number of combinations of 10 objects taken 3 at a time, is

$$C(10,3) = \frac{P(10,3)}{3!} = \frac{10 \cdot 9 \cdot 8}{3 \cdot 2 \cdot 1} = 120.$$

There are 120 ways of choosing 3 of 10 distinct objects without regard to order.

In how many ways can no objects be chosen from n objects? The answer, of course, is 1; just don't choose any, that's the one way. This leads us to define $P(n,0)$ and $C(n,0)$ as follows:

$$P(n,0) = C(n,0) = 1$$

A frequently asked question is, when is order important and when is it not important? When is a situation a permutation situation and when is it a combination situation? We have formulas for computing the number of permutations and combinations inherent in a situation, but there is no formula for deciding which is to be used. This is a matter of judgment which requires a careful reading and analysis of the situation. It is where we are most on our own.

Example 7

How many lines are determined by 12 points, no three of which lie on the same line?

The problem reduces to determining the number of ways of choosing 2 of 12 points without regard to order, which C(12,2) = 66. A line is determined by two points, irrespective of order. (The line determined by points P and Q is the same as the one determined by Q and P.) The condition that no three points lie on the same line is essential to avoiding line duplication. If P, Q and R lie on the same line, P and Q, P and R, and Q and R would determine the same line and 66 would overcount the number of lines determined by the 12 points.

Example 8

The student Math Society at Huxley College has 25 members. An election is to be held to elect a president, secretary and treasurer from its membership. In how many ways can an election slate be formed if no person may hold more than one office?

The problem reduces to determining the number of ways of choosing 3 distinct club members (no repetitions since no one may hold more than one office) with regard to order, which is P(25,3) = 25 · 24 · 23 = 13,800. The order feature is determined by the offices to be filled, which must be distinguished.

Example 9

Two percent of a lot of 100 items are known to be defective. In how many ways can (a) a sample of 3 items be drawn from the lot; (b) a sample of 3 items, all of which are good, be drawn from the lot; (c) a sample of 3, one of which is defective, be drawn from the lot?

(a) Assuming that our interest in the sample is its composition from the point of view of good versus defective items, and not in any order in which the items may appear, the problem reduces to determining the number of ways of choosing 3 of 100 items without regard to order. This is given by C(100,3) = 161,700.

(b) If all items drawn are to be good, they must be drawn from the 98 good ones. There are C(98,3) = 152,096 ways to do this.

(c) To form this sample consider two procedures. The first is to choose 1 of the 2 defectives, which can be done in C(2,1) = 2 ways. The second is to choose the 2 other items needed to make up the sample of 3 from the 98 good ones, which can be done in C(98,2) = 4,753 ways. By the multiplication principle, the number of samples of 3 items containing 1 defective that can be drawn from the lot is C(2,1) · C(98,2) = 9,506.

Food for Thought

17. Evaluate P(7,3), P(12,4), P(6,1), C(16,3), C(18,2) and C(52,4).

18. A club consisting of 60 members meets to elect 4 officers, president, vice president, secretary and treasurer, from its membership. In how many ways can this slate be formed if no person may occupy more than one position?

19. A guest house with 12 single rooms receives 6 single reservations. In how many ways can these reservations be filled?

20. The committee on sabbaticals at Huxley College has authority to grant 5 sabbaticals for any given year. In how many ways can the sabbaticals be granted if 15 requests are received?

21. How many choices of 3 suits and 4 ties for a trip can be made from a wardrobe of 5 suits and 6 ties?

22. There are 4 vacancies on the state Court of Appeals. In how many ways can these vacancies be filled if 20 names have been placed in nomination?

23. How many permutations of the letters of the English alphabet are there?

24. The Alumni Association of Ecap University has organized a one mile race to be run by 2 faculty, W.J. Adams and H. Lurier, and 3 alumni of Ecap University, J. Ross, M. Tilson and E. Kapp. (a) How many possible finishes are there? (b) In how many finishes does Adams finish first? (c) In how many finishes do alumni finish in the first three places?

25. From a lot of 50 color television sets, a sample of 3 is selected for inspection. There are 4 defective sets in the lot. (a) How many samples of 3 of 50 sets are there? How many of these samples contain (b) no defective sets; (c) 1 defective set; (d) 2 defective sets; (e) 3 defective sets?

26. There are 12 faculty in the mathematics department and 10 faculty in the economics department of Ecap University. A joint committee of 5 faculty is to be set up to study curriculum questions of interest to both departments. In how many ways can such a committee be formed if the committee

   a. is to contain 2 members of the mathematics department;

   b. is to contain the chairperson of both departments;

   c. is to contain the chairperson of both departments, but is not to contain Professor Adelson of the mathematics department?

THREE PROBABILITY PROBLEMS

Example 10. The One Mile Race

The setting of this problem is provided by Exercise 24. The Alumni Association of Ecap University has organized a one mile race to be run by 2 faculty, W.J. Adams and H. Lurier, and 3 alumni, J. Ross, M. Tilson and E. Kapp.

What is the probability that Adams finishes first?

One approach to this problem is to note that there are $5! = 120$ possible finishes, that Adams is first in $4! = 24$ of them, and conclude that the probability that Adams finishes first is $4!$ divided by $5!$, or $1/5$.

This approach, which is based solely on counting, leaves much to be desired. It is based on an underlying assumption of equally likely outcomes, but which outcomes are assumed to be equally likely is not made clear and is left to the imagination of the reader. The absence of an explicitly stated probability model and assumption on which the probability function of the model is based obscures the necessity for a critical examination of the realism of the assumption made and suggests the mistaken view that there is only one probabilistic conclusion possible which is an unassailable truth. This kind of approach, which is far too commonly seen in applications of probability, should be accompanied by a skull and cross-bones to warn the reader that his perspective and understanding are in danger of being poisoned.

To analyze the question posed, we will have to back up and provide a probability model for the process along with a justification for the model's probability function, which is open to scrutiny.

As to notation, let (ALRTK), to take an example, denote the outcome indicated by the order, Adams (1st), Lurier (2nd), Ross (3rd), Tilson (4th), Kapp (5th). We take as our sample space S the outcomes expressed by all permutations of A,L,R,T and K. There are $5! = 120$ sample points in S.

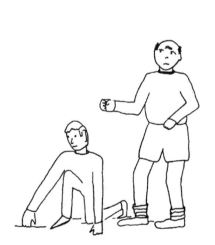

If all five runners are in comparable physical condition, age and running experience, then this would make reasonable an assignment of equal probabilities of 1/120 to the 120 sample points in S, from which it would follow as a valid conclusion that the probability Adams finishes first is 1/5.

Some observers have argued, however, that the five runners are not in comparable physical condition, that Adams tires quickly when the temperature is over 75°F, that the weather forecast is for an 80°F day when the race is to be run, and that it is therefore unrealistic to assign the same probability value to all finishes.

Example 11. Random Sampling

The setting for this problem is provided by Example 9. Two percent of a lot of 100 items are known to be defective. A sample of 3 items is chosen at random from the lot. What is the probability that the sample drawn contains no defectives?

For notational convenience, let us think of the 100 items as tagged I_1, I_2, . . ., I_{100}. Our interest in a chosen sample is in the nature of its items (defective versus good), not in the order in which they are drawn or arranged. This leads us to take as our sample space S the events expressed by all combinations of 100 items taken 3 at a time; that is,

$$S = \{ (I_1, I_2, I_3), \ldots, (I_{98}, I_{99}, I_{100}) \},$$

where (I_1, I_2, I_3), for example, is the event that the sample drawn consists of items I_1, I_2, and I_3. The number of sample points in S is $C(100,3) = 161{,}700$.

The sample is envisioned as being drawn at random, which means in a unbiased manner which does not in any way favor certain samples being drawn over others. The probability function P which best reflects a random drawing in such a situation assigns the same probability value, $1/C(100,3)$ or $1/161{,}700$, to each sample point in S.

$$P(I_1,I_2,I_3) = \ldots P(I_{98},I_{99},I_{100}) = \frac{1}{C(100,3)} = \frac{1}{161,700}$$

In terms of this model, the probability that the sample drawn contains no defectives is:

$$P(no\ defectives) = \frac{C(98,3)}{C(100,3)} = \frac{152,096}{161,700} = 0.941$$

Suppose that our criterion for accepting the entire lot of 100 items from a distributor was that a sample of 3 drawn at random from the lot contain no defectives; then the relative frequency interpretation of our conclusion is that in performing this sampling procedure on lots of 100 items, over the long run we would be accepting such lots with 2% defectives around 94.1% of the time.

Example 12. Population Estimation Problems

How many fish are in your favorite lake? How many raccoons are in your neighborhood? How many animals of your favorite kind are in the game reserve or national park? More generally, how many "whatever" are in your region of interest?

One approach to problems of this sort is based on what is called the **capture—release-recapture** method. We illustrate it by considering a fish population estimation problem, but the approach is applicable to the other situations noted as well.

We begin by catching a certain number of fish from the lake—100, say. These fish are tagged so as to be identifiable if caught again and are thrown back into the lake. We wait for a reasonable time to elapse to allow the fish to disperse (maybe a few days) and then catch another batch of fish, 200, say, and make note of how many in this batch were caught before. Let us suppose that one fish was twice caught.

Let N denote the number of fish in the lake. Our problem is to estimate N. To do this we set up a probability model for the experiment of catching the second batch of 200 fish. As our sample space we take the events represented by the collection of all batches (combinations) of 200 fish that can be selected from N. There are $C(N, 200)$ such batches. Based on the assumption that all fish in the lake have the same likelihood of being caught, we take as our probability function P the one that assigns the same value, $1/C(N, 200)$, to each sample point. We next determine the probability of catching 1 marked fish. This probability is equal to

$$. \quad \frac{\text{number of ways of catching 1 marked fish}}{C(N, 200)} \quad .$$

The number of ways of catching 1 marked fish in a batch of 200 fish is equal to the number of ways of catching 1 of 100 marked fish, $C(100, 1)$, times the number of ways of catching 199 of $N - 100$ unmarked fish, $C(N - 100, 199)$, which yields the product $C(100, 1) \cdot C(N - 100, 199)$. Thus

$$P(1 \text{ marked fish is caught}) = \frac{C(100, 1) \cdot C(N - 100, 199)}{C(N, 200)} \quad (1)$$

The right side of (1) depends on N. It varies as different numbers are substituted for N. Of special interest is that number which when substituted for N makes (1) assume its maximum value. This value is called the maximum likelihood estimate of N; that is, the **maximum likelihood estimate of N** is that number which maximizes the probability of catching the number of marked fish that were actually caught in the second batch.

We shall now show that the maximum likelihood estimate of N is 20,000. The right side of (1) is a function of N, which we shall denote by $P(N)$.

$$P(N) = \frac{C(100, 1) \cdot C(N - 100, 199)}{C(N, 200)} \quad (2)$$

We seek a positive integer value of N such that

$$P(N-1) \le P(N) \quad \text{and} \quad P(N) \ge P(N+1)$$

or, equivalently,

$$\frac{P(N-1)}{P(N)} \le 1 \quad \text{and} \quad \frac{P(N)}{P(N+1)} \ge 1 \qquad (3)$$

Our first task is to determine and simplify $P(N-1)$, $P(N)$, and $P(N+1)$. From (2) we obtain

$$P(N) = \frac{\dfrac{100\,(N-100)\,\ldots\,(N-298)}{199 \cdot 198 \,\ldots\, 1}}{\dfrac{N\,(N-1)\,\ldots\,(N-199)}{200 \cdot 199 \,\ldots\, 1}}$$

$$= \frac{100\,(N-100)\,\ldots\,(N-298)}{199 \cdot 198 \,\ldots\, 1} \cdot \frac{200 \cdot 199 \,\ldots\, 1}{N\,(N-1)\,\ldots\,(N-199)}$$

$$= \frac{100\,(200)(N-100)\,\ldots\,(N-298)}{N(N-1)\,\ldots\,(N-199)} \qquad (4)$$

For $P(N-1)$ we have

$$P(N-1) = \frac{C\,(100,\,1) \cdot C(N-1-100,\,199)}{C(N-1,\,200)}$$

$$= \frac{\dfrac{100\,(N - 101) \ldots (N - 299)}{199 \cdot 198 \ldots 1}}{\dfrac{(N - 1) \ldots (N - 200)}{200 \cdot 199 \ldots 1}}$$

Inverting and simplifying yields

$$P(N - 1) = \frac{100\,(200)(N - 101) \ldots (N - 299)}{(N - 1) \ldots (N - 200)} \qquad (5)$$

For $P(N + 1)$ we have

$$P(N + 1) = \frac{C(100,\ 1) \cdot C(N + 1 - 100,\ 199)}{C(N + 1,\ 200)}$$

$$= \frac{\dfrac{100\,(N - 99) \ldots (N - 297)}{199 \cdot 198 \ldots 1}}{\dfrac{(N + 1) \ldots (N - 198)}{200 \cdot 199 \ldots 1}}$$

Inverting and simplifying yields

$$P(N + 1) = \frac{100\,(200)(N - 99) \ldots (N - 297)}{(N + 1) \ldots (N - 198)} \qquad (6)$$

From (4) and (5) we have

$$\frac{P(N-1)}{P(N)} = \frac{\dfrac{100\,(200\,)(N-101)\,\ldots\,(N-299)}{(N-1)\,\ldots\,(N-200)}}{\dfrac{100\,(200\,)(N-100)\,\ldots\,(N-298)}{N(N-1)\,\ldots\,(N-199)}}$$

Inverting and canceling like terms yields

$$\frac{P(N-1)}{P(N)} = \frac{100\,(200\,)(N-101)\,\ldots\,(N-299)}{(N-1)\,\ldots\,(N-200)} \cdot$$

$$\frac{N(N-1)\,\ldots\,(N-199)}{100\,(200\,)(N-100)\,\ldots\,(N-298)}$$

$$\frac{P(N-1)}{P(N)} = \frac{(N-299\,)N}{(N-200\,)(N-100\,)} \qquad (7)$$

From (4) and (6) we have

$$\frac{P(N)}{P(N+1)} = \frac{\dfrac{100\,(200\,)(N-100)\,\ldots\,(N-298)}{N(N-1)\,\ldots\,(N-199)}}{\dfrac{100\,(200\,)(N-99\,)\,\ldots\,(N-297)}{(N+1)\,\ldots\,(N-198)}}$$

Inverting and canceling like terms yields

$$\frac{P(N)}{P(N+1)} = \frac{100\,(200\,)(N-100)\,\ldots\,(N-298)}{N(N-1)\,\ldots\,(N-199)} \cdot$$

$$\frac{(N+1)\,\ldots\,(N-198)}{100\,(200\,)(N-99\,)\,\ldots\,(N-297)}$$

$$\frac{P(N)}{P(N+1)} = \frac{(N-298)(N+1)}{(N-199)(N-99)} \tag{8}$$

From (3), (7), and (8), our problem reduces to finding N such that

$$\frac{(N-299)N}{(N-200)(N-100)} \le 1 \quad \text{and} \quad \frac{(N-298)(N+1)}{(N-199)(N-99)} \ge 1$$

From the first of these conditions we obtain

$$(N-299)N \le (N-200)(N-100)$$
$$N^2 - 299N \le N^2 - 300N + 20{,}000$$
$$N \le 20{,}000$$

From the second of these conditions we obtain

$$(N-298)(N+1) \ge (N-199)(N-99)$$
$$N^2 - 297N - 298 \ge N^2 - 298N + 19{,}701$$
$$N \ge 19{,}999$$

Thus

$$19{,}999 \le N \le 20{,}000$$

We may take 19,999 or 20,000 as our maximum-likelihood estimate of the size of the fish population.

More generally, the maximum-likelihood function that corresponds to catching k fish from the lake, tagging them and throwing them back, and catching a second batch of n fish that is observed to contain r tagged fish is defined by

$$P(N) = \frac{C(k, r) \cdot C(N - k, n - r)}{C(N, n)}$$

In our example $k = 100$, $n = 200$, and $r = 1$. By an analysis similar to the preceding one, it can be shown that the maximum-likelihood estimate of N is characterized by

$$\frac{nk}{r} - 1 \leq N \leq \frac{nk}{r}$$

Example 13. What is the Raccoon Population?

Judging from complaints about damage caused by raccoons in the community of East Beach, the raccoon population had increased significantly during the last five years along with the human population. But how large was it? This is what the Community Council wanted to find out. They commissioned a team headed by Irving Fine to obtain a maximum likelihood estimate of the raccoon population in East Beach.

Twenty raccoons were caught, tagged, and released. Shortly thereafter 15 were caught and it was found that 3 had been previously caught.

a. Based on these results, what is the maximum likelihood estimate of the raccoon population in East Beach?

b. What assumption(s) underlie this estimate?

c. From the damage her property sustained, Janet Reed felt that the estimate was too low. What aspects of the Fine team's analysis should she look over most carefully in order to satisfy herself that the estimate is realistic, or present a credible case that it is not? Explain.

a. *k*, the number of raccoons initially caught, tagged, and released, is 20; *n*, the number of raccoons caught the second time, is 15; *r*, the number of raccoons twice caught, is 3. Thus, the maximum-likelihood estimate of the raccoon population is:

$$N = \frac{nk}{r} = \frac{15(20)}{3} = 100$$

b. The second group of raccoons caught is a "close approximation" to being a random sample of the raccoon population of East Beach.

c. There are two noteworthy aspects: the first is concerned with the randomness of the second group of 15 raccoons caught. The result obtained is based on the assumption that it was, that is, that there was, to a close approximation, no bias, deliberate or inadvertent, which favored some samples of the raccoon population being caught over others. This is the territory that Janet Reed should look over most carefully. If

the way in which the second sample of raccoons was caught differs "significantly" from being a random sample, then the realism of the Fine team's maximum-likelihood estimate of 100 raccoons would seriously be open to question. If it comes to this, then the 100 value might substantially underestimate or overestimate the raccoon population.

The second noteworthy aspect of this estimate is that even with the random sampling condition being satisfied, it is the best estimate in a probability sense, which is still short of certainty. But then all estimates are short of certainty.

Food for Thought

27. In connection with the relative frequency interpretation of the conclusion obtained in Example 11, suppose that samples of 3 items free of defectives are obtained in 75% of the samples drawn in 500 repetitions of the sampling procedure. How are we to account for the discrepancy between the obtained 75% and the predicted 94.1%?

28. With respect to the probability model described in Example 11, determine the probability that the sample drawn (a) contains 1 defective; (b) contains 2 defectives.

In the following two situations set up a probability model for the process described, state the assumption which underlies the probability function of your model, and determine the probability values required.

29. From a lot of 50 color television sets, a sample of 3 is selected for inspection. There are 4 defective sets in the lot (see Exercise 25). Determine the probability that the sample drawn contains (a) no defective sets; (b) 1 defective set; (c) 2 defective sets.

30. A store's file of 90 accounts contains 12 delinquent accounts and 78 nondelinquent accounts. An auditor plans to choose a sample of 4 accounts for examination. Determine the probability that the sample drawn has (a) no delinquent accounts; (b) 1 delinquent account.

31. Fred Bass caught 9 fish, 3 of which were smaller than the law permits to be caught. A game warden inspects the catch by selecting 2 fish at random from the fisherman's bag and examining them. Some questions of interest to Fred are: What is the probability that no undersized fish are selected? What is the probability that at least one undersized fish is selected? To answer these questions, Fred set up a probability model by taking as a sample space $S = \{f_1, f_2, f_3, f_4, f_5, f_6, f_7, f_8, f_9\}$, where f_1 is the event that fish 1 is selected, f_2 is the event that fish 2 is selected, and so on, and taking as his probability function P the one that assigns the same value, 1/9, to each of the sample points. From this probability model he concluded that the probability that no undersized fish are selected is $C(6,2)C(9,2) = 5/12$, and that the probability that at least one undersized fish is selected is $1 - 5/12 = 7/12$. Is Fred's analysis correct? Explain. How would you analyze the problem?

32. The squirrel population seems to have exploded in Bell City and the City Council would like to obtain an estimate of the size of this population. Darius Consultants were hired to obtain a maximum-likelihood estimate of the population size.

    Fifty squirrels were caught, tagged, and released. Shortly thereafter 40 were caught and it was found that 2 had been previously caught.

    a. Based on these results, what is the maximum-likelihood estimate of the squirrel population in Bell City?

    b. What assumption(s) underlie this estimate?

    c. Is it possible that the estimate is considerably off the mark? Explain.

33. A lot of 20 items is known to contain 2 defectives. Consider an inspection procedure that consists of selecting 2 items at random from the shipment, one after the other, where the first item selected is not replaced before the second one is drawn. Jim Williams was interested in finding the probability of the event B that the sample drawn contains 2 good items, and set up a probability model with sample space $S = \{G_1G_2, G_1D_2, D_1G_2, D_1D_2\}$, where G_1G_2 is the event that the first and second

items drawn are good, G_1D_2 is the event that the first item drawn is good and the second item drawn is defective, and so on. Jim assigned equal probabilities of $\frac{1}{4}$ to these sample points and found the probability to be $\frac{1}{4}$ that the sample drawn contains 2 good items.

a. Is Jim's conclusion valid? Explain.

b. State the relative-frequency interpretation of Jim's conclusion.

c. Take 20 pennies, 2 that are new and shiny (to represent the 2 defective items in the lot) and 18 that have lost their luster (to represent the 18 good items in the lot), put them in a bag, shake the bag, and, without peeking, draw 2 pennies from the bag, one after the other. Repeat the process 300 times, record the occurrence of event B, and find the relative-frequency of B for the 300 repetitions of the process. Compare the result obtained with the relative-frequency interpretation of Jim's conclusion.

d. Does the result obtained affect the validity of Jim's conclusion? Explain.

e. Is Jim's probability model realistic? Explain.

f. How would you set up a probability model for the selection process?

g. Find the probability that the sample drawn contains 2 good items in your probability model, and interpret your result in relative-frequency terms.

h. Do the findings obtained in (c) support the results obtained in (g)? Explain.

34. For the purpose of obtaining a maximum-likelihood estimate of the fish population of Lake Mark, 300 fish were caught, tagged, and released into the lake. Shortly thereafter, 500 fish were caught from the lake, and it was found that 2 had been previously caught.

a. Determine the maximum-likelihood estimate of the fish population.

b. What assumption(s) underlie this estimate?

c. Is it possible that the estimate is much too high? Explain.

d. Is it possible that the estimate is much too low? Explain.

8.6 MARKOV CHAINS FOR MARKETING

THE CONCEPT OF MARKOV CHAIN

To introduce the concept of Markov chain, consider an atomic particle in random motion along a circle (Figure 8.3). Suppose the particle is initially located at one of the four positions, 1 let us say, shown on the circle. We will call this initial position of the particle the **initial state.**

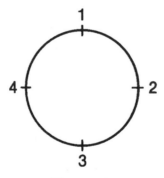

Figure 8.3

Let E_i denote the event that the particle is in position i, more generally called state i. The initial state E_1 of the particle is position 1.

From state E_1 (position 1) the particle can jump clockwise to position 2, or counterclockwise to position 4. This movement is called trial 1; if the particle jumps to position 2, the outcome is state E_2; if it jumps to position 4, the outcome is state E_4. Assuming the particle moved to state E_2 in trial 1, in trial 2 it may move to state E_3 or to state E_1. If it moves to state E_3, then the outcome of trial 2 is E_3.

For any trial, the probability that the particle jumps clockwise to a neighboring state is, let us assume, 3/5; the probability that it jumps counterclockwise to a neighboring state is 2/5.

The particle's behavior has the following characteristics.

1. Each movement (trial) can result in one of n = 4 states: E_1, E_2, E_3, E_4.

2. The state the particle occupies next depends only on the state it is in now, and not on any previous state.

3. The probability with which the particle moves to any state from its current state is known and is independent of the trial number.

The probability that a transition is made from state E_i to state E_j on a trial, denoted by p_{ij}, is called a **transition probability**. The transition probabilities define the behavior of the particle. They are conveniently displayed in a square array called a **transition matrix.** The transition matrix for the particle is:

$$T = \begin{array}{c} \\ E_1 \\ E_2 \\ E_3 \\ E_4 \end{array} \begin{array}{cccc} E_1 & E_2 & E_3 & E_4 \\ \left[\begin{array}{cccc} 0 & 3/5 & 0 & 2/5 \\ 2/5 & 0 & 3/5 & 0 \\ 0 & 2/5 & 0 & 3/5 \\ 3/5 & 0 & 2/5 & 0 \end{array} \right] \end{array}$$

The entry in row 1, column 1, $p_{11} = 0$, is the probability that the particle goes from state E_1 to E_1 on any one trial; more generally, the entry in row i, column j, p_{ij}, is the probability that the particle goes from state i to state j on any one trial.

This system illustrates a system known as an homogeneous Markov chain. More generally, an **homogeneous Markov chain** is a structure consisting of a sequence of trials with the following properties:

1. Each trial leads to one of n outcomes E_1, E_2, . . ., E_n, called **states**.

2. The probability that state E_i occurs on the kth trial depends only on the state in the preceding (k − 1)th trial, and not on the states of the system in earlier trials.

The matrix T of transition probabilities,

$$
T = \begin{array}{c} \\ E_1 \\ E_2 \\ \\ \cdot \\ \cdot \\ E_n \end{array} \begin{bmatrix} E_1 & E_2 & \cdots & E_n \\ p_{11} & p_{12} & \cdots & p_{1n} \\ p_{21} & p_{22} & \cdots & p_{2n} \\ \cdot & \cdot & \cdot & \\ \cdot & \cdot & \cdot & \\ \cdot & \cdot & \cdot & \\ p_{n1} & p_{n2} & \cdots & p_{nn} \end{bmatrix}
$$

describes the basic characteristics of an homogeneous Markov chain.

The Markov chain structure, named for Andrei Andreyevich Markov (1856–1922) who initiated a systematic study of such processes in the early part of this century, has found important applications in physics, biology and business. The application of the Markov chain model to a problem in marketing also provides us with an interesting case study of the danger of forcing a model onto a situation where it is not a good fit.

MARKOV FOR MARKETING?

The late 1950s and 1960s saw the development of Markov chain models to describe consumer brand choice behavior. A starting point for many of these investigations is the view that a brand loyalty and brand switching matrix of probabilities can be constructed from data on sequences of consumer purchases.

$$T_1 = \begin{bmatrix} p_{11} & p_{12} & \cdots & p_{1n} \\ p_{21} & p_{22} & \cdots & p_{2n} \\ \cdot & \cdot & \cdot & \cdot \\ \cdot & \cdot & \cdot & \cdot \\ \cdot & \cdot & \cdot & \cdot \\ p_{n1} & p_{n2} & \cdots & p_{nn} \end{bmatrix}$$

can be constructed from data on sequences of consumer purchases. The value p_{11}, for example, expresses the probability that the consumer, having bought brand 1 in the last period, will also purchase brand 1 in the next period. More generally, p_{ij} expresses the probability that the consumer, having bought brand i in the last period, will purchase brand j in the next period.

An early application of this sort is one undertaken by Benjamin Lipstein [8] concerning the test marketing of a new margarine, fictitiously called Electra, in the Chicago area from November 1958 to May 1959. In Lipstein's study the possible states a margarine buyer could be in were the following:

E_1: Electra Brand E_4: Aunt Mary's brand

E_2: Gloria E_5: Meadowlark brand

E_3: B-R Stores brand E_6: All other brands

E_7: Did not buy margarine during time period

Lipstein's paper contains the brand loyalty and brand switching matrix of transition probabilities (Table 8.4) which represents the situation in the margarine market in Chicago shortly after the introduction of the new brand Electra.

Table 8.4

Next period	Electra	Gloria	B-R	Aunt Mary's	Meadow-lark	Other	Did not buy
Electra	.12	.05	.03	.02	.04	.03	.05
Gloria	.05	.25	.02	.05	.01	.05	.03
B-R	.07	.03	.21	.01	.03	.03.	.04
Aunt Mary's	.04	.02	.05	.23	.02	.04	.01
Meadowlark	.03	.02	.03	.04	.22	.05	.02
Other	.28	.26	.26	.25	.30	.23	.28
Did not buy	.41	.37	.40	.40	.38	.57	.57

But are Markov chain models realistic for the study of consumer brand choice behavior? A critical appraisal was given by A.S.C. Ehrenberg [1], who expressed the view that "frequent public reference to Markov-brand switching models had not been matched by an obvious array of published demonstration of their practical effectiveness." On the basis of a detailed discussion and analysis, Ehrenberg concluded that

the failure of the Markov brand-switching model to live up to its earlier public reputation need not be surprising if seen as an example of misguided but perhaps understandable enthusiasm for forcing an attractively simple piece of college mathematics (stationary Markov theory) onto repeat-buying and brand-switching data while:

1. Omitting to ensure that the data are of a technically suitable form to be modeled by the model.

2. Omitting to examine the crucial assumption involved.

3. Omitting any self-critical appraisal of the various concepts and analytical steps in the approach.

4. Omitting to gather any generalized empirical knowledge of repeat-buying and brand-switching behavior as such.

William F. Massey and Donald G. Morrison [12] expressed agreement with many of Ehrenberg's arguments that the simple Markov chain does not fit all, or even many, real brand-switching situations, but felt that Ehrenberg had been too harsh in his judgment. They expressed the view that the basic Markovian approach is fruitful and should not be abandoned. In a reply [2], Ehrenberg took issue with Massey and Morrison and again raised the question, "Can we not bury Markov for Marketing?"

The Ehrenberg-Massey-Morrison exchange should serve to remind us that mathematical models can only provide us with valid conclusions with respect to our assumptions. If the assumptions are unrealistic, then the mathematical model does not properly fit the situation, and to try to force a fit can be as counterproductive and painful as forcing a pair of size 8 shoes on feet that require size 10.

As to Lipstein's study, it follows from his transition matrix that over the long term Electra brand would end up with about 4% of the margarine market. However, six months or so after its introduction Electra succeeded in capturing about 12% of the market. Electra had been effective in building up the percentage of buyers who having purchased Electra in one period, remained loyal to it in the next period, so that the 0.12 value in the first row, first column of Table 8.4 would have to be revised to 0.23. The need to change transition probabilities had not been taken into account in Lipstein's model.

REFERENCES

1. A.S.C. Ehrenberg. "An Appraisal of Markov Brand-Switching Models," *Journal of Marketing Research*, vol. 2, no. 4 (Nov. 1964), pp. 347–362.

2. ___. "On Clarifying M and M," *Journal of Marketing Research*, vol. 5, no. 2 (May 1968), pp. 228–29.

3. Jean E. Draper and Larry H. Nolan. "A Markov Chain Analysis of Brand Preference." *Journal of Advertising Research*, vol. 4, no. 3 (September 1964), pp. 33–39.

4. Frank Harary and Benjamin Lipstein. "The Dynamics of Brand Loyalty: A Markovian Approach." *Operations Research*, vol. 10., no. 1 (January–February 1962), pp. 19–40.

5. Jerome D. Herniter and John F. Mag. "Customer Behavior as a Markov Process." *Operations Research*, vol. 9, no. 1 (January–February 1961), pp. 105–122.

6. Ronald A. Howard. "Stochastic Process Models of Consumer Behavior." *Journal of Advertising Research*, vol. 3, no. 3 (September 1963), pp. 35–42. Reprinted in *Marketing Models: Quantitative Applications*, edited by R.L. Day and L.J. Parsons, pp. 104–117. Scranton, Pa.: Intext Educational Publishers, 1971.

7. Benjamin Lipstein. "The Dynamics of Brand Loyalty and Brand Switching." *Proceedings of the Fifth Annual Conference of the Advertising Research Foundation* (1959), pp. 101–108.

8. "Tests for Test Marketing," *Harvard Business Review*, vol. 76 (March–April, 1961), pp. 365–369.

9. "A Mathematical Model of Consumer Behavior." *Journal of Marketing Research*, vol. 2, no. 3 (August 1965), pp. 259–265. Reprinted in *Marketing Models: Quantitative Applications*, edited by R.L. Day and L.J. Parsons, pp. 65–79.

10. P.A. Longton and B.T. Warner. "A Mathematical Model for Marketing." *Metra*, vol. 1 (September 1962), pp. 297–310.

11. Richard B. Maffei. "Brand Preferences and Simple Markov Processes." *Operations Research*, vol. 8, no. 2 (March–April 1960), pp. 210–218.

12. William Massey and Donald Morrison. "Comments on Ehrenberg's Appraisal of Brand-Switching Models." *Journal of Marketing Research,* vol. 5, no. 2 (May 1968), pp. 225–227.

13. George P.H. Styan and Harry Smith Jr. "Markov Chains Applied to Marketing." *Journal of Marketing Research*, vol. 1, no. 1 (February 1964), pp. 50–55.

8.7 INTERPRETATIONS OF PROBABILITY

Historically, the study of the stability exhibited by long run relative frequencies of events connected with games of chance played a fundamental role in the birth and early development of the theory of probability. But having been born, probability theory began a separate mathematical life of its own. This mathematical life begins with the concept of probability model, and consists of the theorems and definitions that are built up on the basis of this concept.

Although the relative frequency point of view suggested the formulation of many concepts of the mathematical theory of probability, this interpretation, and any other interpretation for that matter, are not part of the internal structure of this mathematical theory. The relationship between a mathematical theory and envisioned interpretation of the theory that suggested many of the theory's concepts is somewhat similar to the relationship between parents and child. The parents give birth to the child and influence the child's development. Although we may hear comments about physical and temperament similarities between parents and child, we recognize the child as a separate and distinct individual. In a similar sense, a mathematical theory is a structure separate from an envisioned application or interpretation that may have played the role of parents in giving rise to the theory.

Since a given interpretation of a mathematical theory is a structure separate from the theory itself, it might be possible and useful to interpret the theory's concepts in other ways. Such is the case with probability theory. Another interpretation that has received much attention in recent years is one in which the probability value assigned to an event is interpreted as a quantitative measure of an individual's degree of belief in the occurrence of the event. An ardent New York Mets fan might say that the Mets have a 90 percent chance of taking the pennant next year; translation: the probability that the Mets take the pennant next year is 0.90. A less ardent Mets admirer might say that the Mets have a 10 percent chance of taking the pennant next year. A business man might say that there's an 80 percent chance that the sales volume of the firm will top $5 million this year. Individuals often have beliefs or opinions about possible outcomes connected with situations in which the outcome is not certain. Such an individual sometimes finds it useful to assign a value between 0 and 1, inclusive, to a possible outcome as a quantitative measure of his feelings about the likelihood of occurrence of the outcome. A strong opinion about the occurrence of an event is reflected by the assignment of a value close to 1 to the event; a strong opinion about the nonoccurrence of an event is reflected by the assignment of a value close to 0 to the event. Probability values that are assigned to an event from this point of view, or that are interpreted in this way, are called **personal** or **subjective probabilities** and are said to express a person's degree of belief in the occurrence of the event.

The point of view of subjective probability admits as meaningful probabilistic assertions about events connected with situations that occur once and cannot be repeated; the relative frequency interpretation, on the other hand, is only meaningful for random processes that can be repeated a large number of times. The subjective point of view admits as meaningful such statements as the probability that the Smith Company will spend more than $2 million on advertising this year is 0.8, and the probability that Jack will not study for his math exam is 0.9. Such statements are meaningless from the relative frequency point of view. An important feature, as well as difficulty, of subjective probability is that the assignment of subjective probabilities to outcomes depends very much on the person doing the assigning. Two individuals might assign markedly different subjective probabilities to the same event.

Subjective probability has been criticized along the following lines: if mathematical probability is regarded as a quantitative measure of a person's degree of belief, then the theory of probability is somewhat like a branch of psychology. The final result of a purely subjective interpretation of probability is subjective idealism. To assume that the evaluation of probability only concerns the feelings of the observer implies that conclusions based on probabilistic judgments are deprived of the objective meaning that they have independent of the observer. In defense of subjective probability, its proponents argue that there are many important once-and-only situations, especially in business, where a decision must be reached. By using subjective probabilities to express his judgments (based on data, analysis, experience, etc.) in quantitative terms, the decision maker can employ the machinery of the mathematical theory of probability to arrive at conclusions that are valid with respect to his judgments translated into quantitative terms. It is understood that the decision maker is not arriving at truth. But he does obtain conclusions that are consistent with his judgments, and such conclusions are often helpful for making the decision he is responsible for.

The interpretation of probabilities as measures of degrees of belief remains controversial at this point. Rather than take sides for or against the subjective point of view, it is more important for us to obtain as thorough an understanding of this point of view as possible and to be able to recognize the subjective use of probability when it appears.

Example 1

A.M. Bradley, who is suffering from a rare disease, was told by his friend that, since recent medical statistics show that 75 percent of those who have had the disease recovered, the probability that he will recover is 0.75. How is this probability assignment to Bradley's recovery to be interpreted?

Although a relative frequency background is involved (recent medical statistics), the focus is on a once-and-only situation, A.M. Bradley's state of health. This by itself is sufficient to exclude a relative frequency interpretation; to repeat the process a large number of times would entail giving Bradley the disease a large number of times and observing how often he recovers, a procedure that does have its difficulties. Bradley's friend is using background relative frequency data as a basis for expressing in quantitative terms his degree of belief in Bradley's recovery.

The following example is due to my colleague Irwin Kabus.

Example 2. Subjective Probability in Banking

For the last twenty years, the top management of Morgan Guaranty Trust Company has been using a technique called histogramming to quantify and picture the uncertainty that surrounds future interest rates on which the bank's asset/liability decisions are based.

One utilizes the histogram technique by first listing all the possible outcomes he believes may result from some future situation and then by assigning subjective probabilities to each of these outcomes. From observing the width of the range of outcomes and the chance associated with each a reader of the histogram is able to assess how confident the histogram maker feels about his judgments. For example, an individual's feeling about the probable level of the interest rate on a 90-day CD at some future date—say, three months from now—might look like the one shown in Figure 8.4.

Each of the interest rates shown represents an interval extending 1/8% below and 1/8% above it. The percentages of subjective chances (show in the vertical bars) are those that the individual has chosen to spread over the possible rates.

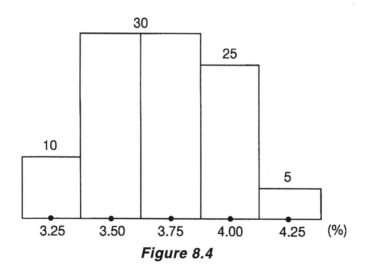

Figure 8.4

Mathematically, this histogram states that the individual feels there is a 10% chance that in three months the 90-day CD rate will be between 3 and 1/8% and 3 and 3/8%, a 30% chance that it will be between 3 and 3/8% and 3 and 5/8%, a 30% chance it will be between 3 and 5/8% and 3 and 7/8%, a 25% chance it will be between 3 and 7/8% and 4 and 1/8%, and a 5% chance it will be between 4 and 1/8% and 4 and 3/8%. However, and this is of the utmost importance, the histogram is not meant to be read with mathematical precision; rather, it is meant to convey a message in terms that express judgments in numerical form.

Qualitatively, this histogram indicates that the individual feels very confident that in three months the 90-day CD rate will be between 3 and 3/8% and 4 and 1/8%. However, he is very uncertain as to what value, within the range, the rate will take on. There are small chances that the rate could drop to 3 and 1/8% to 3 and 3/8% or rise as high as 4 and 1/8% to 4 and 3/8%, although it is more likely that the lower rate is attained than the higher one.

As is most often the case, top management does not base its decisions solely on the opinion of one individual but on the opinions of several. Thus, it becomes necessary to produce an histogram whose qualitative message is some combination of the thinking of all those individuals involved in the process. Individual histograms can be combined to form a single weighted average histogram, with the weights being assigned by the individual responsible for endorsing the final histogram that top management sees.

In cases where all individual histograms receive equal weights (as is true at Morgan Guaranty) the weighted average becomes a simple arithmetic average. In other cases where individual histograms are weighted differently, the weights are generally based on the qualifications and track records of the individuals producing the histograms.

At Morgan Guaranty the histogram committee consisted of seven key executives from various areas of the bank. After collecting each individual's histogram, a summary sheet was generated which collectively showed the subjective probabilities assigned by each committee member. This summary served as the basis for a discussion (moderated by the analyst coordinating the histogram process) in which those members with differing views, immediately identifiable from the summary sheet, had a chance to defend them. At the end of the discussion the members of the histogram committee had the opportunity to change their histograms if they felt inclined to do so after hearing other opinions. The final histograms were then averaged into one consensus histogram which represented the views of the committee as a whole and was then presented to top management. When the moderator of the committee meeting presented the results to top management it was the qualitative story that was given to them.

REFERENCES

1. I. Kabus, "You can Bank on Uncertainty," *Harvard Business Review*, 54,3 (May–June 1976), 95–105.

2. L.J. Savage, *The Foundations of Statistics* (New York: Wiley and Sons, Inc., June 1976), 95–105.

3. R. Schlaifer, *Probability and Statistics for Business Decisions* (New York: McGraw Hill, 1959).

4. R. Schlaifer, *Analysis of Decisions Under Uncertainty* (New York: McGraw Hill, 1969).

Food for Thought

1. It is asserted that the probability of an event E is 0.80. State the relative frequency and subjective probability interpretations of this assertion, and describe the main features of these interpretations.

2. After watching a pair of dice being tossed 100 times, an observer commented that the probability of an even sum showing on the 101st toss is 0.85. Is this probability assignment one that is to be interpreted in relative frequency or subjective probability terms? How so?

3. The following comment appeared in an article on natural gas supplies (*The New York Times*, Feb. 22, 1977, p. 14): "How much

gas is left to be discovered? . . . The last Geological Survey estimate, made two years ago, was this: Given available technology and current economics, there is a 95 percent probability that 322,000 billion cubic feet can be located and produced; there is a 5 percent probability that 655,000 billion cubic feet can be located and produced." How should these probabilistic statements be interpreted? How so?

4. Eric Roberts, a student at Huxley College, commented that the probability that he will get an A in Sociology this semester is 0.95. Is this probability value to be interpreted in relative frequency or subjective probability terms. How so?

5. In a letter to the editor of *The New York Times* (Feb. 28, 1971) on the background of the atomic bomb, Hans Bethe wrote, "By February 1945 it appeared to me and to other fully informed scientists that there was a better than 90 per cent probability that the atomic bomb would in fact explode . . . "How should this probabilistic statement be interpreted? How so?

6. On Dec. 29, 1978 two acoustics experts said tests showed a probability of 95% or better that a shot was fired from a grassy knoll in Dallas when President John F. Kennedy was assassinated. This testimony was presented before the House Select Committee on assassinations (*The New York Times*, Dec. 30, 1978). How should this probability statement be interpreted?

A QUESTION OF MATHEMATICAL PRECISION

9.1 FOOD FOR THOUGHT

1. While on your way to your Aunt Alice's birthday celebration, her hundredth rumor has it, the world seemed to be in conspiracy against you. It was snowing, traffic was bumper to bumper, and then someone in an old oldsmobile sideswiped your new Buick and took off. You finally arrive at your aunt's celebration, but, needless to say, not in very good humor. Before you know it you find yourself in a social group being "educated" by Uncle Harry. Harry thinks he knows everything about everything and on this occasion his subject is mathematics, specifically geometry. "Since their truth was established by the precise mathematical reasoning for which the ancient Greek mathematicians are justly famous, the truth of the theorems of Euclidean geometry is beyond question," bellows Harry in his most authoritative sounding tone. Usually your attitude toward Harry is one of toleration, but this time you're ready for bear. What reply would give to his "profound observation?"

2. The following point of view was expressed at an economics seminar. Would you agree? "Mathematical methods in economics have the advantage of certitude. No qualified person can resist the truth of a mathematical conclusion properly communicated. The job of communication may be difficult if the solution is complex, but when the communication is competent, agreement is inevitable. If anyone doubts a solution, he can recalculate the equations and check the steps in the derivation. Then he must

either demonstrate that there has been an error or acknowledge the truth of the solution."

3. For his birthday, Herman received a pair of dice which, he was told, were well-balanced. On the basis of the probability model that assigns the same value, 1/36, to each of the sample points in the well-known 36-element sample space for the dice tossing process, Herman determined the probability of an even sum showing to be 0.50. He expected that an even sum would show in the neighborhood of 500 times for 1000 tosses of his dice and made betting plans accordingly. Herman participated in a friendly game one evening, and after 1000 tosses of his dice an even sum had showed 200 times and Herman was $600 poorer. Disappointed, confused, and angry, Herman raised the following questions.

   a. If mathematics is such a precise subject, how could this happen?

   b. Isn't my conclusion correct?

   c. What went wrong?

4. Two new lamp models, L-10 and L-15, are to be put into production by Turin Lamps, Inc. A linear programming model was set up by a consulting firm to find the number of lamps of each kind that should be made per week to maximize profit. The solution, obtained by the corner point method, yielded 2500 L-10 lamps and 3000 L-15 lamps. The production manager told the board of directors that since a precise mathematical technique had been used to obtain the solution, the company could feel confident about the accuracy of the result and proceed to set the production schedule at 2500 L-10 and 3000 L-15 lamps per week to maximize profit.

   If you were on the board of directors, would this be a sufficient basis for you to set the production schedule as described?

5. The following conversation took place between Kelly, a girl with a probabilistic problem, and her friend Cathy.

   Kelly took a pair of dice from her bag and asked her friend Cathy to help her determine the probability that the sum of the numbers showing is odd. Cathy was happy to help; she pointed out that the probability that an odd sum shows is 18/36 since the dice can fall in 36 ways and that in 18 of these outcomes the sum of the numbers showing is odd. Cathy further assured her friend that since this conclusion had been obtained by mathematical reasoning, which is very precise, she could have confidence in its truth. Moreover, should a suitable gaming occasion arise, you should not hesitate to act on this fact, said Cathy to Kelly.

   Do you agree with Cathy's analysis and advice? Explain. What reply would you give to Kelly?

10 ▶ THE MATHEMATICS OF SPACE

10.1 FOOD FOR THOUGHT

Carefully consider the following statements. If you agree with a statement, explain the basis for your agreement; if you disagree with a statement, explain the basis for your disagreement.

1. The parallel postulate problem was to show that Euclid's parallel postulate is true.

2. Lobachevsky sought to prove that Euclidean geometry is inconsistent.

3. Immanuel Kant's philosophical views on the nature of mathematics provided a favorable climate for the development of non-Euclidean geometry.

4. Since Lobachevskian geometry contains statements which contradict statements in Euclidean geometry, Lobachevskian geometry cannot be considered a realistic description of physical space.

5. If Euclidean geometry is not a perfect description of space, then the proofs of some Euclidean theorems must be in error.

6. Eugenio Beltrami showed that if Lobachevskian geometry is consistent, then Euclidean geometry must be inconsistent.

7. Janos Bolyai's main achievement was to show that Euclid's parallel postulate is false.

8. A postulate is a statement which is true.

9. "Karl Friedrich Gauss, as a youth, sought to prove that parallel lines will never meet. His failure to do so led him to suspect that there is something imprecise about the geometry of Euclid that is taught in school. It seemed that if parallel lines were extended far enough, they might curve enough to meet." ("Finding of Blue Galaxies Backs Big Bang Theory," *The New York Times*, June 13, 1965).

10. The *reducio ad absurdum* or indirect approach to the parallel postulate problem was to show that the negation of Euclid's parallel postulate is false.

11. In geometry truth and validity mean the same thing.

12. The modern concept of geometry is the same as that advanced by the ancient Greek mathematicians.

13. To say that the postulates of a geometric system are false is a contradiction to the very meaning of a geometric system.

14. Although Lobachevskian geometry is internally free of contradictions, it is not a realistic description of physical space.

15. If the theorems of a geometric system are true, then its postulates must be true since true theorems can only be obtained from true postulates.

16. The validity of mathematical proof in geometry is dependent on the realism of the postulates of the system.

17. A mathematician proposed a geometry based on the postulates of Euclidean geometry with two replacements:

Euclidean postulate	**Replacement**
1. Parallel postulate: Given a line L and point P not on it, there is one line passing through P parallel to L.	1. There are no parallel lines.
2. Given two distinct points, there is one line containing them	2. Given two distinct points, there is at least one line (possibly more than one) containing them.

One argument brought against this proposed geometry is that it must contain contradictory statements (i.e., be inconsistent) because its parallel postulate (there are no parallel lines) and its second statement (two distinct points do not determine one and only one line) are contrary to the nature of physical space. Would you agree with this point of view?

18. To prove a statement P in geometry by *reducio ad absurdum*, we prove its negation not-P to be a valid consequence of the postulates of geometry.

19. One of the most important roles of proof in geometry is that it enables us to show that the postulates of geometry are true or false.

20. The study of non-Euclidean geometry brings nothing to students but fatigue, vanity, arrogance, and imbecility. Non-Euclidean

space is the false invention of demons who gladly furnish the dark understandings of the non-Euclideans with false knowledge.

21. "The most suggestive and notable achievement of the last century is the discovery of Non-Euclidean Geometry." David Hilbert.

MATHEMATICAL MODELING AS A TOOL FOR INQUIRY

11.1 FOOD FOR THOUGHT

1. Apart from "I need another drink," would you agree with the views expressed by the seminar participants pictured below? Explain.

2. "A refined radar technique that may settle the current debate over the validity of Einstein's general theory of relativity has been successfully tested." (*The New York Times*, Feb. 28, 1968, p.20.)

   a. Could such a technique be used to settle a question of validity?

   b. What issue might such a technique help to resolve?

3. "The most accurate long-distance measurements ever made, by means of radio signals between the Viking spacecraft on Mars and antennas on Earth, have produced new confirmation of Einstein's theory of relativity, a Viking project scientist reported today." (*The New York Times*, Jan. 7, 1977, p. A8.) In what sense was the theory of relativity confirmed?

4. "Since its development in 1947, the theory of quantum electrodynamics has enabled physicists to make accurate predictions about the interaction of atomic particles and to develop important electronic technology based on these interactions. But a team of scientists at the University of Michigan has uncovered evidence that the theory may be fundamentally flawed. The group reported in *Physical Review Letters* that atoms of a bizarre, short-lived substance called positronium annihilate themselves significantly faster than the theory of quantum electrodynamics predicts, and hence, there may be something seriously wrong with the theory." (*The New York Times*, April 7, 1987, p. C-4.)

   a. In what sense may there be something seriously wrong with the theory?

   b. In principle, what has happened here?

5. "Japanese scientists have reported that small gyroscopes lose weight when spun under certain conditions. . . . Dr. Park of the University of Maryland said the finding, if proved true, would almost certainly be explained by general relativity, Albert Einstein's geometric theory of gravitation. . . . Dr. Forward, who aids the Air Force in its propulsion work, said the sheer volume of anti-gravity claims threw doubt on the validity of the new finding." (*The New York Times*, December 28, 1989, p. A17.)

a. In what sense might general relativity explain the gyroscope phenomenon?

b. If the gyroscope phenomenon holds up and it cannot be explained by general relativity, would implications would this have for general relativity?

c. In what sense does the "sheer volume of anti-gravity claims" throw doubt on the validity of the new gyroscope finding?

6. "In the past decade scientists have devised a theoretical arrangement known as the standard model, which purportedly shows the fundamental particles of all matter are related. . . . All the particles in the model except the top quark have been discovered." (*The New York Times*, June 13, 1989, p. 6.)

a. Suppose it is shown that the top quark does not exist; what implications would this have for the standard model?

b. Suppose the top quark is found; what implications would this have for the standard model?

7. The Asta Company makes shoes. The Company plans to introduce two new styles, designated by A-18 and A-21, for the fall season and management wants to know how to set its monthly production schedule so as to maximize profit. Two consulting firms were hired to study this problem and to make recommendations. Each consulting firm formulated a mathematical model, designated by M1 and M2, for this problem. A conclusion obtained from M1 states that to maximize profit 50,000 A-18 pairs and 35,000 A-21 pairs should be produced monthly with an anticipated monthly profit of $200,000. A conclusion obtained from M2 states that to maximize profit 40,000 A-18 and 50,000 A-21 pairs should be produced monthly with an anticipated profit of $230,000.

As a member of the board of directors of the Asta Company you are presented with these findings. What questions would you ask?

8. "Evidence reported by physicists last fall suggested that the particles [protons], the basic building blocks of matter, had long, but finite lifetimes. But a recent report by researchers who par-

ticipated in an Ohio study says that the proton lifetimes may be even longer than the billions of years previously estimated. They also say this could mean theories predicting such decay are invalid." (*The New York Times*, Jan. 23, 1983).

a. Would this evidence render the conclusions of such theories of proton decay invalid?

b. What effect would this evidence have on theories of proton decay?

9. What would you say in response to the following view?

Newton devised a theory of planetary motions which scientists believed was true for over two hundred years. How could we have confidence in the theories of science when a theory believed in for so long turned out to be false?

10. ". . . theorists [Albert Einstein and Satyendra Nath Bose] calcu-lated [on the basis of quantum theory] that if a certain class of atoms could be chilled to temperatures below any that exist in nature, the atoms would merge with each other to become hugh 'superatoms:' . . . The creation of a Bose-Einstein condensate, as this hypothetical superatomic state of matter is called, . . . would not only demonstrate the validity of some outlandish predictions of quantum theory, but would create a form of matter that may never have existed anywhere before." (M. Browne, "Physicists Get Warmer in Search for Weird Matter Close to Absolute Zero," *The New York Times*, Aug. 23, 1994, C1.)

a. Would the creation of a Bose-Einstein condensate demon-strate the validity of some outlandish predictions of quantum theory? Explain.

b. What would the creation of a Bose-Einstein condensate dem-onstrate? Explain.

c. In July of 1995 two teams of physicists, working inde-pendently, announced success in creating the sought after Bose-Einstein condensate. See, for example, P. Spotts, "Sci-entists Create New Form of Matter," *The Christian Science Monitor*, July 14, 1995, p. 1; and M. Browne, "2 Groups of

Physicists Produce Matter that Einstein Postulated," *The New York Times*, July 14, 1995, p. A1.

   i. Spotts writes: ". . . researchers appear to have proven a theory first proposed by Albert Einstein and Indian physicist Satyendra Nath Bose 70 years ago." What "theory" was proved and in what sense?

   ii. Is the term "postulated" employed in the title of Browne's article appropriate? Did Bose and Einstein postulate the existence of what was subsequently called a Bose-Einstein condensate?

11. Describe a scientific theory (or model) whose predictions were subsequently confirmed by experimentation or observation.

12. Update on the standard model introduced in 6. In April of 1994 an international team of scientists working at the Fermi National Accelerator Laboratory in Batavia, Illinois, announced the most reliable sightings obtained to date of the top quark.

   a. One view has it that this discovery "is a highly intellectual achievement because the standard model, which it appears to validate, is central to understanding the nature of time, matter and the universe." (*The New York Times*, April 26, 1994 p. A1.)

   b. Another view is that "the quest begun by philosophers in Ancient Greece may have ended . . . with the discovery of evidence for the top quark." (*Ibid.*)

   c. Another view has it that "the Fermilab discovery, if confirmed, would be a major milestone for modern physics because it would complete the experimental proof of . . . the standard model, which defines the modern understanding of the atom and its structure." (*Ibid.*)

   d. Still another view is that "if the top quark could not be found, the standard model of theoretical physicists would collapse, touching off an intellectual crises that would force scientists to rethink three decades of work." (*Ibid.*)

Would you agree with these views? Explain.

13. A scientist proposed the idea that atoms have the structure of a miniature solar system. He cannot check this hypothesis directly, for atoms are invisible to the most powerful microscope.

    a. How can he use mathematics to check his idea?

    b. Can mathematics tell him that he is absolutely right? Explain.

    c. Can mathematics tell him that he is wrong? Explain.

    d. Can mathematics tell him that he is approximately right? Explain.

14. "Imaginary universes are so much more beautiful than this stupidly constructed real one; and most of the finest products of an applied mathematician's fancy must be rejected, as soon as they have been created, for the brutal but sufficient reason that they do not fit the facts." (G.H. Hardy, *A Mathematician's Apology*; Cambridge, 1940). What is Hardy talking about? What are his imaginary universes? In what sense must most of the finest products of a mathematician's fancy be rejected?

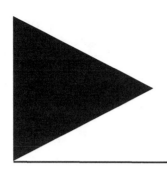

COMMENTS AND ANSWERS TO SELECTED FOOD FOR THOUGHT QUESTIONS

Section 1.1 (page 3)

1. No; John's cancellation of $(x - 2)$ is allowable for x other than 2; when x is 2 we have a division by 0 situation and division by 0 is not and cannot be defined. It is meaningless for John to cancel $(x - 2)$ under the provision that x is not 2, and then proceed to substitute 2 for x.

2. No   3. No   4. No

Section 1.2 (page 3)

1. (a) 1.9 yrs.,   (b) 3.3 yrs.,   (c) 3.8 yrs.,   (d) 3.2; 15.9; 0.8 yrs.

3. (a) 1078 yrs., (b) c. 915 A.D. Development of feudalism in Western Europe, 800–1300; the Fatimid caliphate is established in North Africa; Tollan, in the Valley of Mexico, becomes the capital of the Toltecs, 900–1200.

5. $100 trillion translates to roughly 3.2 million years; $126 trillion translates to roughly 4 million years. The human ancestor Australopithecus afarensis, considered the common root of the human tree, flourished from 3.9 to 3.0 million years ago.

7. 100 quadrillion calculations translates to roughly 3200 million or 3.2 billion years, which exceeds the aforenoted 3.2 million year value by a factor of 1000. The emergence of the first primitive life forms occurred around 3500 million years ago. This was followed by the emergence of single celled organisms, similar algae, which was the main form of life for about 3000 million years. Around 570 million years ago more complex organisms began to evolve. Between 200 and 65 million years ago the earth was populated by dinosaurs.

Section 3.3 (page 21)

1. (a) 510,   (b) 29,524,   (c) 16,383,   (d) 13.5

3. $S_{52} = 2^{52} - 1$, which is of the order of \$4.5 billion.

Section 3.4 (page 30)

1. (a) 10% per annum, (b) (i) 1 year, (ii) \$105, (iii) \$605.

3. (a) \$4105.71, (b) \$4118.37

5. \$4973.76

7. The present value of \$150 is $150(1.02)^{-2} = \$144.18$. The selling price is \$200 + \$144.18 = \$344.18.

Section 3.4 (pages 35,41)

9. $F = 8000 \, S_{\overline{30}| \, 0.05} = \$531,510$

11. (a) \$226.73, (b) \$116.75, (c) \$58.03

13. \$9245.91

15. \$3010.75

17. The cost of the car, $850, is equal to the downpayment D plus the present value P of an annuity with R = $40, n = 12, and i = 0.01; P = $450.20. From 850 = D + 450.20, we obtain D = $399.80.

19. (a) $1347.78, (b) $667.98, (c) $332.52

21. (a) F = $10,000, n = 20, and i = 0.03. Thus R =

    $F \cdot \dfrac{1}{S_{\overline{20}|\,0.03}} = 10{,}000(0.03722) = \$372.20$. Thus $372.20 must

    be deposited into the fund at the end of each 6-month period to accumulate to $10,000 by the time Baxter is 65 years old.

    (b) P = $10,000, n = 20, i = 0.03. Thus R =

    $P \cdot \dfrac{1}{a_{\overline{20}|\,0.03}} = 10{,}000(0.06722) = \$672.20$. Therefore, upon

    retirement he will receive $672.20 at the end of each 6-month period.

23. We first find the rent that she is paying. P $2000, n = 24, i = 0.01.

    Thus $R = P \cdot \dfrac{1}{a_{\overline{24}|\,0.01}} = 2000(0.04707) = \$94.14$. Therefore at the

    end of 15 months she will have paid $94.14(15) = $1412.25 in principal and interest.

    We now determine the interest she will have paid after 15 payments. $2000 at 12% per annum compounded monthly grows to $2000(1.01)^{15} = \$2321.94$. Thus the interest accumulated is $2321.94 − $2000 = $321.94. At the end of 15 months she will have paid $321.94 in interest.

    Thus the principal paid after 15 payments is $1412.25 − $321.94 = $1090.31.

    Therefore the amount of principal remaining to be paid after 15 months is $2000 − $1090.31 = $909.69.

25. If payments are made at the beginning of each payment period, then we obtain as the future value F of the annuity

$$F = R(1 + i) + R(1 + i)^2 + \dots + R(1 + i)^n \qquad (1)$$

If payments are made at the end of each payment period, then the future value of the annuity is expressed by

$$R + R(1 + i) + ... + R(1 + i)^{n-1} = R \cdot s_{\overline{n}\,i} \qquad (2)$$

By multiplying both sides of (2) by $(1 + i)$ we obtain

$$R(1 + i) + R(1 + i)^2 + ... + R(1 + i)^n = (1 + i)R \cdot s_{\overline{n}\,i} \qquad (3)$$

From (1) let us observe that the left side of (3) is F. Thus we have

$$F = (1 + i)R \cdot s_{\overline{n}\,i}$$

By means of a similar analysis it can be shown that the present value P of the annuity for which payments are made at the beginning of each payment period is expressed by

$$P = (1 + i)R \cdot a_{\overline{n}\,i}$$

Section 4.1 (page 51)

1. (a) No. The data, simply interpreted, might give us a measure of the "efficiency" or "productivity" of each instructor for each class, but this does not translate to a measure of the over-all "efficiency" or "productivity" of the department.

    (b) No. This game of academic musical chairs will not change the department's over-all tuition revenue or costs, on which financial efficiency depends.

3. Not by itself; you would also like to take into account the number of departures and the number of miles flown.

5. Not by itself; there are gems, junk, and a lot in between. The "quality" of the research should, in some way, be taken into account.

8. a. Paired data giving the instructional cost and number of students taking the course would have to be obtained for each course.

   b. The instruction cost for Professor White for ENG 205 is $84,000/7 = $12,000. Tuition revenue is $1000 (16) = $16,000. ENG 205 would not run since R − C = $4000 does not exceed $5000.

   c. 18 students.

   d. Yes: C = $6000, R = $16,000 and R − C = $10,000 exceeds $5000.

   e. Very vulnerable.

   f. Not too well. It addresses the cost dimension in a simplistic manner which makes the running or cancellation of a section highly vulnerable to faculty program changes. It does not support academic quality; Professor White is a recognized authority on Modern American Poetry and it may be that Professor Roberts is not as well equipped to handle this course. It's less expensive to have him do this, of course, but insight and expertise are not equivalent for faculty.

   The data needed to implement HA-1 are the paired data described in answer to (a) and involve a significant amount of resources to obtain. Every time there is a program change the data has to be updated for the sections involved.

9. a. The salary of each department member is needed so that the average salary for that department can be determined. The number of students in each course would be needed.

   b. Yes; C = $10,000, R = $16,000, and R − C = $6000, which exceeds $5000.

c. As to merits, HA-2 is an improvement over HA-1 in that the run/cancel decision for a course is not dependent on the instructor assigned to the course but on the number of students enrolled in it, which makes more sense. HA-2 is clearly easier to implement than HA-1.

d. As to disadvantages, if the department consisted mostly of relatively high paid senior faculty with an average salary of $105,000, for example, an enrollment of 20 in a course would be needed to satisfy HA-2. If the department consisted mostly of recently hired junior faculty with an average salary of $35,000, let us assume, an enrollment of 10 in a course would be needed to satisfy HA-2. The implementation of HA-2 might vary considerably from department to department depending on the mix of the department's junior and senior faculty. This does not make good academic sense. Departments are not unconnected units to be considered in isolation, but are intended to service a larger whole—the university itself. The course run/cancel criteria should reflect the needs of the university as a whole, and the shortcoming of HA-2 is that is it not designed to do this.

e. The data required for the implementation of HA-2 is simple to obtain. The shortcoming in the data is due to the shortcoming of HA-2 in reflecting the academic needs of the university, and therefore the students, as a whole.

Section 5.1 (page 61)

2. No; *The Literary Digest's* pre-presidential poll of 1936 makes this very clear. See *Get a Grip on Your Math*, Chapter 8, Section 2.

3. Disagree. The "random" sample was limited to only home owners and thus was not representative of the entire public.

4. No. He only heard from people who cared enough about the issue or poll to call. The rest of the voters could possibly turn the tide.

5. "You got lucky." The mistakes made in the 1932 poll did not surface, but they did in the 1936 poll.

6. They made assumptions on how the undecided would vote and they were wrong. They also assumed that public opinion would remain relatively stable from the time of their last polls in September to election day in November, which proved to be wrong.

7. Pro: Getting insightful comments from the correct population, i.e., guests who actually stayed at the hotel.

    Con: Only getting responses from those who cared to fill out the questionnaire, which is not representative of the general population of guests.

9. They were worded poorly. They contained biases that might strongly influence responses. The yes/no option strongly limits respondent choices of replies.

12. In posing questions Andy might want to strike a balance between manageability (which favors short answer questions) and insightfulness (which favors explanatory type questions). Andy might consider questions of the following type.

    1. How would you rate your overall educational experience at Huxley? (1) Excellent, (2) Good, (3) Satisfactory, (4) Poor, (5) Very poor.

    2. In terms of your major area of study, how would rate your educational experience at Huxley? (1) Excellent, (2) Good, (3) Satisfactory, (4) Poor, (5) Very poor.

    3. If you could go back to the time that you were choosing a school, would you still choose Huxley knowing what you now know? (a) Yes, (b) Probably yes, (c) No, (d) Probably not, (e) Not sure.

    4. What do you view as Huxley's strong points?

    5. What do you view as Huxley's weak points?

    6. What has been your most positive educational experience at Huxley thus far?

7. What has been your most negative educational experience at Huxley thus far?

8. How might Huxley improve its educational quality?

9. Would you recommend Huxley to a close friend or relative? (a) Yes, (b) No, (c) Not sure.

10. Please briefly explain the basis for your answer to the preceding question.

13. a. Would you like Huxley College to provide an exercise room?

How often would you use an exercise room?

Would you be willing to contribute some of your own time to help maintain the exercise room?

(Similar questions for the other facilities.)

b. Greatly desire, Moderately desire, Indifferent, Moderately dislike, Greatly dislike.

Section 5.2 (page 76)

1. a. Probability sampling methods include probability tools so that an estimate of the precision of the results can be made. Nonprobability methods do not contain these probability tools.

b. No. If a measure of the error in the sampling results is not required, then nonprobability sampling might be satisfactory. In general, nonprobability sampling is cheaper and easier to carry out.

4. (i) One might want to do a study to determine the average salary paid to college presidents in the United States. A sample of 100 presidents is to be taken. One approach is to randomly choose two colleges from each of the 50 states *(strata)* to insure that all parts of the country were included in the study.

(ii) One might want to do a study of the differences between salaries of upper, middle, and lower management in corporations. In order to do this, random selections are made from these three management levels *(strata)* in the companies used for the study.

5. The setting described in Question 5 might prompt Andy to introduce *strata* based on student characteristics which might influence student facility preferences for the new student union building. One way of defining *strata* is according to gender, age (less than 30, 30 or over, for example), and school division affiliation (Arts and Sciences, Business, Education, Mathematics and Statistics, Equine Studies), which defines 20 strata.

If Andy planned to conduct personal interviews, complicated by the fact that Huxley is a live-on-campus college with student residences spread over a considerable area, he might want to give cluster sampling serious consideration with clusters defined by the numbers identifying student residences. He might proceed by choosing a sample of student residences and interviewing all or a sample of the students who live there. Cluster sampling under such geographic conditions might reduce considerably the travel time needed to conduct the interviews when random sampling or stratified sampling are employed. Time, as we appreciate, is a precious resource when one is faced with tight deadlines.

If a comprehensive list of students attending Huxley with suitable information on how to contact them were not available, random sampling would not be feasible. Cluster sampling might be the only viable option within the probability sampling framework, assuming that a list of student residences could be obtained from the housing office.

7. No. The make-up of the busiest street is not necessarily (and probably not) the same as the make-up of all streets in the city.

8. No. For a sense of general faculty opinion a random sample should be chosen from the faculty. Here we do only the Business School.

9. No. Using only the opinions of CEO's will almost certainly not represent the opinion of the whole community which is made up of all types of business people.

10. No. It will only give the opinion of those that voted and this may have a bias in it.

11. If the tickets were thoroughly mixed so that each ticket in the bowl had the same chance of being selected, then the dean is right. However, if this is not the case, then Reynolds is correct.

Section 6.1 (page 87)

1. Valid; the hypothesis forces Joe Warren into the category of frogs.

3. C1 is not valid since it is not forced by the hypothesis. C2, C3, and C4 are not valid.

5. C1 is valid.

8. C1 is valid; C2 is not valid.

9. C1: valid; C2: valid; C3: not valid; C4: not valid; C5: not valid; C6: not valid.

Section 6.2 (page 93)

1. The problem here is that what was actually proved is not the same as what was required to be proved. The proof given showed that a unique line PE parallel to L is obtained by the argument developed. This does not show that a line M parallel to L obtained from some other construction must be the same as PE. Indeed, other constructions of parallel lines to L at P are known.

2. More than the three cases cited are possible and would have to be considered. Consider, for example, the following possibility: angle 1 + angle 4 > 180° and angle 2 + angle 3 < 180°. This does not contradict the relationship that the sum of the four angles equals 360°.

Section 6.3 (page 103)

1. **Model 1.** Zogs: ordered pairs (points) (1,1), (2,1), (2,2), (1,2); Glob: the set determined by any two of the points given (see Figure 1).

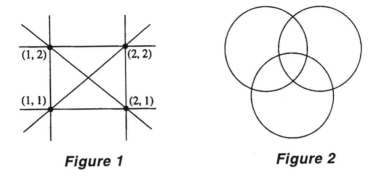

Figure 1   **Figure 2**

Model 2. Zogs: points inside three overlapping circles that intersect (the globs) as shown in Figure 2.

3. No; the argument does not establish that s and q are distinct.

5. No; Jane is in trouble with her first statement which introduces four zogs based on Conjecture 1 that there must be four zogs in Glob Theory. Conjecture 1 has not been proved a theorem and, in fact, is not a theorem since models of Glob Theory with three zogs have been exhibited.

6. No; that the argument given in 3 is invalid does not by itself exclude the possibility that there is a valid argument establishing 3.

9. (a) **Model 1:** Euclidean plane geometry obtained by interpreting luk and yuk as point and line in the usual way.

    Model 2: Luks: the points with coordinates $(2,0)$, $(0,2)$, $(-2,0)$, and $(0,-2)$ (see Figure 3). Yuks: sets of 2 of these points. There are 6 yuks as indicated in Figure 1: $L_1 = \{(2,0), (0,2)\}$, $L_2 = \{(0,2), (-2,0)\}$, $L_3 = \{(-2,0), (0,-2)\}$, $L_4 = \{(0,-2), (2,0)\}$, $L_5 = \{(0,2), (0,-2)\}$, $L_6 = \{(-2,0), (2,0)\}$.

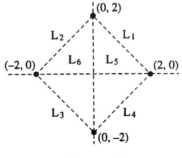

Figure 3

(b)–(h) We run into difficulty when attempting to construct a model for Yuk Theory
with fewer than four luks and
six yuks. This by itself is not
conclusive since it may simply
mean that we've not been
clever enough and there are
such models, but it does suggest that C2 – C6 are theorems,
which they are. The construction of Model 2 with six yuks
shows that C7 is not a theorem.
As to C1, it holds in Model 2,
but this is not conclusive for

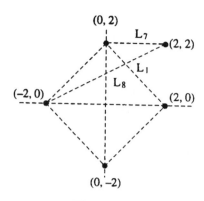

Figure 4

the general case. We might try to play a bit with Model 2 and come
up with a point which is contained by only one set of two points.
Consider (2,2), for example, and define $L_7 = \{(0,2), (2,2)\}$ (see
Figure 4). If we add these to Model 2 and leave it at that, the modified
system does not satisfy P5. (For L_1 and (2,2), there is no L_i containing
(2,2) which is parallel to L_1). We might try to close this gap by
introducing $L_8 = \{(2,2), (-2,0)\}$; but then we have a violation of P5
from another point of view. (For L_8 and (0,2) there are two luks, L_1
and L_5, containing (0,2) which are parallel to L_8). After a while with
this sort of unsuccessful playing around, it begins to seem that C1
might be a theorem. It is.

Theorem 1. Every luk is contained by at least two yuks.

Statement	Justification
1. Let P denote any luk in the system. We shall show the p is contained by at least two yuks.	1. P2.
2. Let q denote another luk in the system.	2. P2.
3. Let L_1 denote a yuk containing p and q.	3. P3.
4. Let r denote a luk not contained by L_1.	4. P4.
5. Let L_2 denote a yuk containing p and r.	5. P3.

In summary, we have that p, a "representative" luk in the system, is contained by distinct yuks L_1 and L_2; L_2 differs from L_1 because it contains r which is not contained by L_1.

Section 6.4 (page 111)

1. No; C1, "there are at least three neighborhoods in Neighborhood Theory," is a theorem which contradicts P4.

3. Model 1 in answer to 9 (a) shows that Yuk Theory is as consistent as Euclidean geometry.

4. No; P6 contradicts C6, there are at least six yuks, which is a theorem in the system.

5. The question reduces to this: Is there a model for postulates P1–P4 for which P5 is not satisfied? If so, then the answer is yes, P5 is independent. Let us go back to Model 2 of Yuk Theory given in answer to 9(a) and add to it the luk (2,2) and yuks L_7 = {(2,2), (0,2)}, L_8 = {(2,2), (-2,0)}, L_9 = {(2,2), (0,-2)}, and L_{10} = {(2,2), (2,0)} (see Figure 5). This system satisfies postulates P1–P4 of Yuk Theory, but not P5. For yuk L_8 and luk (0,2) there are two yuks, L_1 and L_5 containing (0,2) which are parallel to L_8.

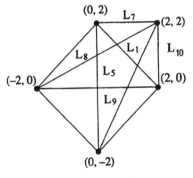

Figure 5

Section 6.5 (page 114)

1. a. Disagree,   b. Disagree,   c. Agree

   d. Disagree,   e. Disagree,   f. Disagree

   g. Disagree,   h. Disagree.

3. The model's assumptions (or postulates).

Section 7.2 (pages 127, 132)

1. (25,350)
2. (40,179)
3. (50,355)
4. (245,50)
5. (3,3)
6. (3/2,3/2)
7. (1,2)
8. (2,1)

9. (45,15)
10. (150,000; 100,000)
11. (100,300)
12. (120,250)
13. (50,40)
14. (60,15)
15. (20,80)

21. See Figure 6
24. See Figure 8
27. See Figure 10

23. See Figure 7
25. See Figure 9

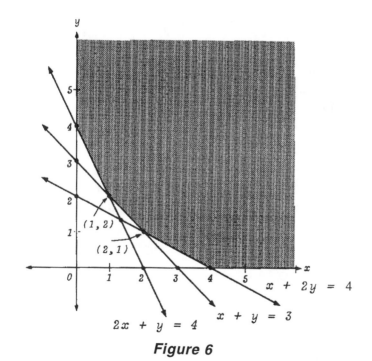

Figure 6

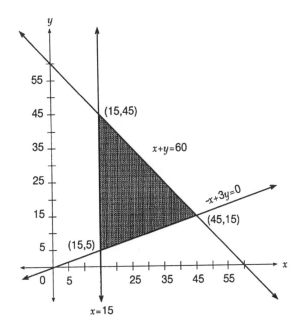

Figure 7

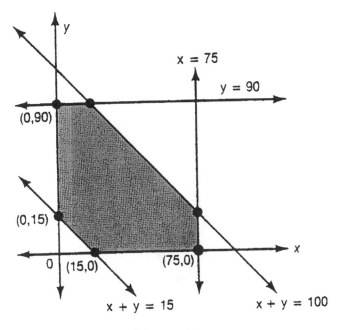

Figure 8

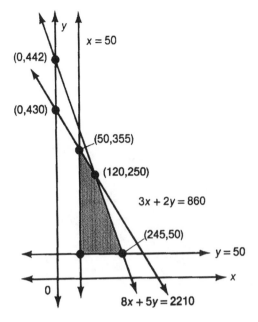

Figure 9

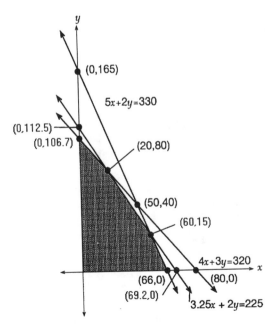

Figure 10

Section 7.3 (page 139)

1. The corner-point method cannot by itself guarantee that profit will be maximized when 300 ZKB-47 and 250 ZKB-82 units are made daily and sold. Whether these output levels or other ones will maximize profit is a question of truth, and the corner-point method, as a mathematical technique, can only ensure the validity of the predicted output levels with respect to the linear programming model that was set up for the profit maximization problem in question. If the assumptions that the model reflects are realistic, then the aforementioned output levels, obtained as a valid conclusion of these assumptions, will maximize profit; if these assumptions are not realistic, then it might well happen that other output levels would yield a higher profit than that projected for 300 ZKB-47 and 250 ZKB-82 units.

   The basis for implementing the conclusion that 300 ZKB-47 and 250 ZKB-82 units be made and sold daily is the belief, based on an analysis of the company's operations and the market, that the assumptions made are realistic.

2. a. No, mathematics is precise in the sense that its methods yield a solution which is valid with respect to the linear program in question and the assumptions that it reflects. Since different assumptions were made, reflected by different linear programs, there are two solutions, each of which is valid with respect to its underlying linear program.

   b. No; different valid conclusions are not in conflict in the sense of validity because they come from different models.

   c. What assumptions underlie your linear program models? Convince me that they are realistic. What features were considered negligible and are not reflected in the models? What is the basis for treating these features as negligible? There is one constraint, $3.25x + 2y \leq 225$, found in LP-2 which is not found in LP-1; what is the basis for its inclusion in LP-2 and absence in LP-1? The objective functions of LP-1 and LP-2 are different; what is the basis for this difference?

   d. I would implement the conclusion obtained from the model whose assumptions I view as realistic. If I had fundamental reservations about the realism of the assumptions of both models, I would not adopt either model.

3. Corner points: (0,0), (0,5), (3,3), (6,0). Solution: (6,0); max. value: 30.

5. The graph of the feasible points is shown in Figure 6 in answer to Question 21 of Section 7.2. The corner points are (0,4), (1,2), (2,1), and (4,0). Solution: (4,0); max value: 6.

6. The graph of the feasible points is shown in Figure 7 in answer to Question 23 of Section 7.2. The corner points are (15,5), (15,45), and (45,15).

8. The graph of the feasible points is shown in Figure 8 in answer to Question 24 of Section 7.2. The corner points are (15,0), (75,0), (75,25), (10,90), (0,90), and (0,15). The corner points (75,0) and (75,25) are solutions yielding the minimum value 12,420. Non-corner points on the line segment joining (75,0) and (75,25) are solutions as well, yielding 12,420.

The corner point theorem does not exclude the possibility of non-corner point solutions. It guarantees a solution from the corner points, provided that the linear program has a solution to begin with.

9. The graph of the feasible points is shown in Figure 9 in answer to Question 25 of Section 7.2. Solution: (50,355); max. value: 60,250.

11. The graph of the feasible points is shown in Figure 10 in answer to Question 27 of Section 7.2. Solution: (60,15); max. value: 13,050.

Section 7.5 (page 159)

3. Let x and y denote the number of ounces of orange juice and apricot juice concentrate, respectively, to make up a container of fruit juice.

Min C = 3x + 2y
subject to
$$x \geq 0, y \geq 0$$
$$2x + 3y \geq 120$$
$$3x + 2y \geq 150$$
$$x + y \geq 55$$

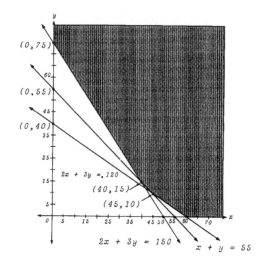

Figure 11

The graph of the feasible points is shown in Figure 11. Solutions: (0,75), (40,15) and all points (x,y) on the line segment joining them; min. value: 150.

5. Let x and y denote the number of associate and full-professor slots, respectively, to be established. The Faculty Council's objective function to be maximized is F(x,y) = x + y while the administration's objective function to be minimized is C(x,y) = 5000x + 10,000y. The constraints for both objective functions are:

$$x \geq 0, y \geq 0$$
$$x \leq 22, y \geq 3$$
$$x + 2y \leq 30$$
$$-x + 4y \leq 0$$

The Faculty Council's linear program has solution (22,4) and maximum value 26. Implementation of this conclusion calls for establishing 22 associate professor slots and 4 full-professor slots. The administration's linear program has solution (12,3) and minimum value 90000. Implementation of this conclusion calls for establishing 12 slots at the associate professor level and 3 slots at the full-professor level. One should always have concerns about the realism of assumptions being made, but there are more conditions of a straightforward nature (for example, at least 3 full-professor slots must be established) than assumptions of a more controversial nature. The one assumption that might warrant more examination is that "at most $150,000 is available for merit increments." Why a ceiling of $150,000 rather than $120,000 or $200,000?

6.   Let x denote the amount (in millions of dollars) allocated for loans, and let y denote the amount (in millions of dollars) allocated for securities. The investment income function to be maximized is

$$I(x, y) = 0.09x + 0.06y$$

The inequality

$$x + y \leq 25$$

expresses the condition that a total of $25 million is available for investment in loans and securities.

The inequality

$$y \geq \frac{1}{5}(x + y)$$

or equivalently,

$$-x + 4y \geq 0$$

expresses the condition that a securities balance equal to or greater than 20 percent of total assets must be maintained.

The requirement that at least $8 million must be available for loans is expressed by the inequality

$$x \geq 8$$

We thus obtain the following linear program:

$$\text{Max } I(x, y) = 0.09x + 0.06y$$

subject to
$$x \geq 0$$
$$y \geq 0$$
$$x + y \leq 25$$
$$-x + 4y \geq 0$$
$$x \geq 8$$

The graph of the feasible points is shown in Figure 12. The corner points are $(8,2)$, $(20,5)$ and $(8,17)$, with $(20,5)$ yielding the maximum value 2.1.

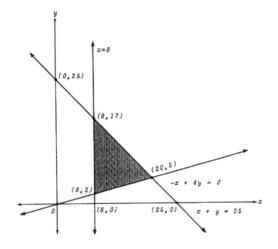

Figure 12

8. Let x and y denote the number of hours per day that the Brooks and Darius mines, respectively, are operated. Basic data are summarized in Table 1.

Table 1

	Number of hours in operation	Cost per hour	High-grade ore per hour (tons)	Medium-grade ore per hour (tons)
Brooks	x	$500	1	4
Darius	y	$700	2	3

The following linear program emerges:

Min C(x,y) = 500x + 700y

subject to

x ≥ 0, y ≥ 0

x + 2y ≥ 40:   At least 40 tons of high-grade ore are needed per day.

4x + 3y ≥ 100: At least 100 tons of medium-grade ore are needed per day.

x ≤ 24, y ≤ 24: Neither mine can operate more than 24 hours per day.

Solution (16,12); min. value 16,400.

10. Let x and y denote the target sales volume (in product units) for hospitals and medical supply houses, respectively. The linear program that emerges,

$$\text{Max } P(x,y) = 100x + 120y$$

subject to

$$x \geq 0, \, y \geq 0$$
$$x + 2y \leq 24{,}000$$
$$2x + y \leq 30{,}000$$
$$x \geq 5{,}000, \, y \geq 5000,$$

has solution (12000,6000) with maximum value 1,920,000. Implementation of this finding calls for selling 12,000 X-ray units to hospitals and 6000 units to medical supply houses. The assumptions underlying the linear program model developed should be carefully reviewed as a matter of standard procedure, but there are no special concerns that stand out.

11. Let x and y denote the number of coats and dresses to be made, respectively. The linear program that emerges,

$$\text{Max } I(x,y) = 110x + 70y$$

subject to

$$x \geq 0, \, y \geq 0$$
$$x \leq 25$$
$$2x + 4y \leq 120$$
$$4x + 2y \leq 110,$$

has solution (50/3, 65/3) with maximum value 3350. Since the company will have difficulty selling 50/3 coats and 65/3 dresses, this solution cannot be implemented. What is required is a solution in integers which maximizes I(x,y). Such problems are called integer programs. Sometimes the optimal solution obtained by linear program solution methods is a solution in integers, but sometimes it is not. When it is not and a solution in integers is required, then integer programming solution techniques which yield the optimal solution in integers must be employed. The optimal solution in integers to the Hoffman Company's integer program is (17,21), which yields the maximum value 3340. For details see: W. J Adams, *Finite Mathematics, Models, and Structure* (Dubuque: Kendall/Hunt Pub. Co., 1995), Sec. 3.9.

12. Let x and y denote the number of tons of steel produced subject to the F14 and F24 filter systems, respectively. The linear program that emerges,

$$\text{Min } C(x,y) = 1.2x + 1.8y$$

subject to

$$x \geq 0, y \geq 0$$
$$x + y = 2{,}000{,}000$$
$$30x + 32y \geq 62{,}400{,}000,$$

has solution $x = 800{,}000$, $y = 1{,}200{,}000$ with a minimum value of 3,120,000.

15. Let x and y denote the number of B4 and B9 buses to be purchased, respectively. The linear program that emerges,

$$\text{Max } C = 39{,}600x + 54{,}000y$$

subject to

$$x \geq 0, y \geq 0$$
$$2x + 3y \leq 50$$
$$x + y \geq 30,$$

has no solution because the constraints are incompatible. There are times when one can't have it all, and this is one of those times.

16. To relate candidates 1 (Jones), 2 (Johnson), and 3 (Marks) to jobs 1 (Supreme Court Judge) and 2 (Civil Court Judge) we introduce variables X_{11}, X_{21}, and more generally X_{ij}, to relate candidate i to job j, as summarized in Table 2. X_{ij} can assume one of two values, 0 if candidate i is not assigned job j, 1 if candidate i is assigned job j.

Table 2

Candidate	Job	
	Supreme Court Judge (job1)	Civil Court Judge (job 2)
Jones (candidate 1)	X_{11}	X_{12}
Johnson (candidate 2)	X_{21}	X_{22}
Marks (candidate 3)	X_{31}	X_{32}

The problem of filling the positions so that the total potential rating is maximized in such a way that each candidate is assigned to at most one job and each job is filled by at most one person is expressed by the following linear program which requires a solution in integers (0's and 1's).

Maximize $P = 9X_{11} + 8X_{21} + 10X_{31} + 8X_{12} + 9X_{22} + 8X_{32}$

subject to

$$
\begin{aligned}
X_{11} + X_{12} &\le 1 \\
X_{21} + X_{22} &\le 1 \\
X_{31} + X_{32} &\le 1 \\
X_{11} + X_{21} + X_{31} &\le 1 \\
X_{12} + X_{22} + X_{32} &\le 1)
\end{aligned}
$$

Each candidate is assigned to at most one job

Each job is filled by at most one candidate

18. To relate the blood supply to the patients, variables are introduced as shown in Table 3.

Table 3

Blood Type		Patient	
		Levy 1	Rudd 2
A	1	X_{11}	X_{12}
B	2		X_{22}
O	3	X_{31}	X_{32}

The variable X_{ij} represents the number of pints of blood type i to be sent to patient j, assuming that type i is compatible with the blood type of patient j. The following linear program (integer program) emerges:

$$\text{Minimize } C = 30X_{11} + 30X_{12} + 25X_{22} + 20X_{31} + 20X_{32}$$

subject to

$$X_{11} \geq 0, \ X_{12} \geq 0, \text{ etc.}$$

$$
\begin{array}{llll}
X_{11} + & X_{12} & & \leq 2 \\
 & X_{22} & & \leq 3 \\
 & & X_{31} + X_{32} & \leq 2 \\
X_{11} & & + X_{31} & \geq 3 \\
 & X_{12} + X_{22} & + X_{32} & \geq 2
\end{array}
$$

The amount of each blood type cannot exceed the supply.

Each patient must be sent at least the amount required of compatible blood.

Section 8.1 (page 177)

3. S, unconventional as it might seem, is a sample space for the dice tossing process because it has the required property that whenever the dice are tossed, one of the events in S occurs.

6. $S_1 = \{G,D\}$, where D is the event that a defective bulb is selected, and G is the event that a good bulb is selected. $S_2 = \{GP1, GP2, DP1, DP2\}$, where GP1 is the event that a good bulb made by plant P1 is selected, etc. If we think of the bulbs as identified 1, ..., 8000, then $S_3 = \{b_1, b_2, ..., b_{8000}\}$, where b_1 is the event that bulb 1 is chosen, etc., is sample space.

Section 8.3 (page 183)

1. a. $P(E) = P(2) + P(4) + P(6) = 3/4 = 0.75$.

   b. If a die whose behavior is described by model Y3 is tossed a large number of times, an even number will show approximately 75% of the time.

c. The tossing of the die a large number of times and observing how often an even number shows is irrelevant to the validity issue. The validity of $P(E) = 0.75$ was established by adding up the probabilities with which 2, 4 and 6 show in model Y3 and obtaining 0.75.

d. From tossing the die a large number of times we have that the relative frequency with which an even number showed is 0.665, which is markedly at variance with the predicted relative frequency of approximately 0.75 in (b). This establishes that $P(E) = 0.75$, interpreted in relative frequency terms, is false.

e. Model Y3 is not realistic for the die in question since a valid conclusion of the model has been shown to be false.

Section 8.4 (page 190)

1. a. For well-balanced dice it is unrealistic to assume that the sample points in the main diagonal of S_3, $\{1,1\}$, $\{2,2\}$, $\{3,3\}$, etc., are as likely to occur as the ones above it, $\{1,2\}$, $\{1,3\}$, $\{2,3\}$, etc. $\{1,2\}$, for example, can occur in two ways, with the red die showing 1 and green showing 2 as well as with red showing 2 and green showing 1. The sample points in the main diagonal can only occur in one way; $\{1,1\}$, for example, occurs when red shows 1 and green shows 1.

b. Let x denote the value assigned to the sample points in the main diagonal; assign 2x to the others. There are 6 sample points in the main diagonal and 15 others. This yields:

$$6x + 15(2x) = 1$$
$$36x = 1$$
$$x = \frac{1}{36}$$

Assign probability 1/36 to the sample points in the main diagonal and 1/18 to the others.

2. $P(2) = \dfrac{1}{36}$, $P(3) = \dfrac{2}{36}$, $P(4) = \dfrac{3}{36}$, $P(5) = \dfrac{4}{36}$, $P(6) = \dfrac{5}{36}$,

$P(7) = \dfrac{6}{36}$, $P(8) = \dfrac{5}{36}$, $P(9) = \dfrac{4}{36}$, $P(10) = \dfrac{3}{36}$, $P(11) = \dfrac{2}{36}$,

$P(12) = \dfrac{1}{36}$

3. a. (1) Mark's probability function is not realistic because it does not take into account the different proportions of good bulbs made by plant P1, good bulbs made by plant P2, etc. From the data given the proportion of good bulbs (of the total output of 8000 bulbs) made by P1 is 4950/8000, the proportion of defective bulbs made by P1 is 50/8000, etc. (2) Mark's conclusion is valid with respect to his model, but not realistic in terms of the process of choosing a bulb at random from the day's output with the given proportions of good and defective bulbs.

b. Bob's model is open to the same sort of criticism leveled at his brother's model.

c. Let us think of the bulbs as identified 1, 2, ..., 8000. We take as our sample space $S = \{b_1, b_2, ..., b_{8000}\}$, where b_1 is the event bulb 1 is chosen, b_2 is the event bulb 2 is chosen, etc. Since we envision the bulb as being chosen at random, without bias, from the day's output, we take as our probability function P the one which assigns 1/8000 to each sample point in S.

d. Yes, as follows: P(GP1) = 4950/8000, P(GP2) = 2985/8000, P(DP1) = 50/8000, P(DP2) = 15/8000.

e. Yes, P(G) = 7935/8000, P(D) = 65/8000.

Section 8.5 (page 196)

1. There are two places to be filled. First place can be filled in any one of 6 ways, and after it has been filled in one of these ways, second place can be filled in any one of the 5 remaining ways. Thus, by the multiplication principle, the number of ways of filling both places is $6 \cdot 5 = 30$.

2. a. $7 \cdot 6 \cdot 5 \cdot 4 \cdot 3 \cdot 2 \cdot 1 = 5040$

    b. 5040, the total number of arrangements, minus 1, the number of in-order arrangements, leaves us with 5039 out of order arrangements.

    c. In $1 \cdot 1 \cdot 3 \cdot 2 \cdot 1 = 6$ arrangements volume 1 will occupy first place and volume 2 will occupy second place.

3. $5 \cdot 3 = 15$

4. $2 \cdot 4 = 8$

5. $3 \cdot 4 \cdot 4 \cdot 3 = 144$

6. $5 \cdot 4 \cdot 3 \cdot 2 \cdot 1 = 120$

7. $9 \cdot 10 \cdot 10 \cdot 10 = 9000$

8. In terms of the multiplication principle, the trick to this problem is to see to count the number of integers between 1000 and 4999, inclusive. The number of integers between 1000 and 4999 inclusive is $4 \cdot 10 \cdot 10 \cdot 10 = 4000$. (There are four places to be filled and first place can be filled in four ways, with 1, 2, 3, or 4; second place can be filled in ten ways, and so on.) We must add 1 to this total of 4000 to account for the number 5000. Thus we obtain 4001 as the number of integers between 4000 and 5000, inclusive.

    Alternately, there are 5001 integers between 0 and 5000, inclusive, and there are 1000 integers between 0 and 999, inclusive. This leaves $5001 - 1000 = 4001$ integers between 1000 and 5000, inclusive.

9. 2^8, assuming that each question is marked T or F. If we allow blank spaces as an option, then there are 3^8 possible answer sheets.

10. (a) $7 \cdot 6 \cdot 5 \cdot 4 \cdot 3 \cdot 2 \cdot 1 = 5040$, (b) $4 \cdot 3 \cdot 5 \cdot 4 \cdot 3 \cdot 2 \cdot 1 = 1440$
    (c) $4 \cdot 3 \cdot 2 \cdot 1 \cdot 3 \cdot 2 \cdot 1 = 144$, (d) $144 + 144 = 288$

11. There are n places to be filled. First place can be filled with any of the n people and thus in n ways. After first place has been

filled, second place can be filled with any of the remaining n – 1 people and thus in n – 1 ways. After the first two places have been filled, third place can be filled with any of the remaining n – 2 people and thus in n – 2 ways. And so on. By the multiplication principle the number of ways in which the n places can be filled is n(n – 1) ...1.

12. 8 + 6 = 14

13. 2 · 2 · 2 = 8

14. a. 4 · 4 · 3 · 3 · 2 · 2 · 1 · 1 = 576,   b. 576 + 576 = 1152

15. a. 10 · 9 · 8 · 7 · 6 · 5 · 4 = 604,800,   b. 10^7

16. a. 8 · 7 · 6 · 5 · 4 · 3 · 2 · 1 = 40,320,

    b. 9 · 8 · 7 · 6 · 5 · 4 · 3 · 2 · 1 = 362,880,

    c. 4 · 3 · 2 · 1 · 5 · 4 · 3 · 2 · 1 = 2880

Section 8.5 (page 203)

18. P(60,4) = 60 · 59 · 58 · 57 = 11,703,240; why P(60,4) and not C(60,4)?

19. P(12,6) or C(12,6), depending on whether 6 of 12 rooms are to be selected with regard to location, in which case order is important, or not.

20. C(15,5)   21. C(5,3) · C(6,4)   22. C(20,4)

23. 26!   24. (a) 5! = 120,   (b) 1 · 4! = 24,   (c) 3 · 2 · 1 · 2 · 1 = 12

25. a. C(50,3) unordered samples,   b. C(46,3),
    c. C(4,1) · C(46,2),   d. C(4,2) · C(46,1),
    e. C(4,3)

26. a. C(12,2) · C(10,3),   b. C(2,2) · C(20,3) = C(20,3),
    c. C(2,2) · C(19,3) = C(19,3)

Section 8.5 (page 215)

27. The basic condition underlying the model, random sampling, was not lived up to in the manner that the samples were drawn.

28.
a. $P(1 \text{ def.}) = \dfrac{C(2,1) \cdot C(98,2)}{C(100,3)}$,   b. $P(2 \text{ def.}) = \dfrac{C(2,2) \cdot C(98,1)}{C(100,3)}$

29. S is determined by all unordered samples of 3 television sets that can be drawn from the lot; $S = \{(T_1, T_2, T_3), \ldots, (T_{48}, T_{49}, T_{50})\}$. Based on the assumption that the sample is drawn at random from the lot, we are led to the probability function P which assigns $1/C(50,3)$ to each sample point in S.

a. $\dfrac{C(46,3)}{C(50,3)}$,   b. $\dfrac{C(4,1) \cdot C(46,2)}{C(50,3)}$,   c. $\dfrac{C(4,2) \cdot C(46,1)}{C(50,3)}$

30. S is determined by all unordered samples of 4 accounts than can be chosen from the 90 accounts; $S = \{(A_1, A_2, A_3, A_4), \ldots, (A_{87}, A_{88}, A_{89}, A_{90})\}$. Based on the assumption that the sample is drawn at random from the 90 accounts, we are led to the probability function P which assigns $1/C(90,4)$ to each sample point in S.

a. $\dfrac{C(78,4)}{C(90,4)}$,   b. $\dfrac{C(12,1) \cdot C(78,2)}{C(90,4)}$

31. Fred Bass made two major mistakes in his analysis. The first is that the model he set up is not suitable for the process of choosing 2 fish from his bag; it is suitable for the process of choosing 1 fish from the bag. Fred then compounded his error by obtaining a conclusion about the probability that no undersized fish are selected which is not valid with respect to the model that he did set up—a mathematical switching of horses, or should we say fish, in midstream, so-to-speak.

The appropriate way to proceed is to take as the sample space S the events described by all combinations of 9 fish taken 2 at a time.

$$S = \{(f_1,f_2), (f_1,f_3), \ldots, (f_8,f_9)\}$$

There are $C(9,2) = 36$ sample points in S. Based on the assumed randomness of the selection, define P by assigning $1/C(9,2) = 1/36$ to each sample point in S. In terms of this model the probability that no undersized fish are selected is $C(6,2)/C(9,2) = 5/12$.

32. a. $k = 50$, $n = 40$, $r = 2$; $N = 1000$

    b. The second group of squirrels caught is a "close approxima-tion" to being a random sample of the squirrel population.

    c. Yes; if the sample actually drawn deviated significantly from the random sample assumed in the model.

33. a. Jim Williams's conclusion is valid in that it follows as an inescapable consequence of his probability model. $P(B) = P(G_1,G_2) = 1/4$.

    b. If the sampling procedure is repeated a large number of times—that is, two items are drawn at random from a lot of twenty where the first item drawn is not replaced before the second is drawn, and this is done a large number of times— then the sample drawn will consist of two good items approxi-mately 25% of the time.

    c. In performing the underlying process 300 times we have $RF(B) = 238/300 = 0.793$. Event B was found to occur 79.3% of the time, which differs markedly from the 25% value predicted from Mr. Williams's model.

    d. As established in part (a), Mr. Williams's conclusion that the probability that the sample drawn consists of two good items is 0.25 follows as an inescapable consequence of his prob-ability model, and thus is valid with respect to his model by virtue of the meaning of validity. The data cited in part (c) has no bearing on the validity of his conclusion.

e. The result R(B) = 0.793 establishes that Jim's valid conclusion, interpreted in relative frequency terms, is false about the underlying process. This in turn tells us that Jim's model is not a realistic one for the inspection procedure.

f. For convenience of discussion, let us think of the items in the lot as labeled 1, 2, ..., 20. Another approach to the study of the sampling process is to take as our sample space S the events expressed by all combinations of 2 items that can be drawn from the lot of 20 items.

$$S = \{(I_1, I_2), (I_1, I_3), ..., (I_{19}, I_{20})\}$$

The probability function P that best reflects the randomness of the selection procedure is the one that assigns the same value, $1/C(20,2)$ = 1/190, to each sample point in S.

g. From this model we have: $P(2 \text{ good items}) = C(18,2)/C(20,2)$ = 0.805. The relative frequency interpretation of this is that if the sampling procedure is repeated a large number of times, the sample drawn will contain two good items approximately 80.5% of the time.

h. The 79.3% value obtained by repeating the sampling procedure a large number of times is in close agreement with the 80.5% value obtained as a valid consequence of the probability model developed in (f). This provides support for the realism of this model for the sampling procedure.

34. a.
$$N = \frac{300\,(500)}{2} = 75,000$$

b. The second group of fish caught is a "close approximation" to being a random sample of the fish population.

c., d. Yes; if the sample actually caught deviated significantly from the random sample assumed in the model.

Section 8.7 (page 230)

1. Relative-frequency interpretation: if the underlying process is repeated a large number of times, event *E* will occur in the neighborhood of 80% of the time. Subjective interpretation: 0.80 is a numerical expression of some individual's degree of belief in the occurrence of event *E* in connection with the underlying process. The relative-frequency interpretation presupposes that the process can be repeated a large number of times and has an objective content that is independent of the observer. The subjective interpretation can be entertained in connection with once-and-only situations. The subjective probability assigned to an event often depends very much on the observer.

3. Both probabilistic statements can only be interpreted in subjective terms. Both are numerical expressions of degree of belief connected with once-and-only situations.

5. The underlying situation is clearly a once-and-only kind of situation. The probability value cited is a quantitative measure of Dr. Bethe's degree of belief that the atomic bomb that had been constructed would explode.

Section 9.1 (page 233)

1. Disagree; precise mathematical reasoning in whatever context, geometric or other, can only establish the validity of the conclusions in question based on the assumptions or postulates set up as a starting point. The truth of the conclusions is another issue altogether.

2. No; truth and validity are being equated here. It would be appropriate to say that, no qualified person can resist the *validity* of a mathematical conclusion. In the last sentence of the statement again replace truth by validity.

3. a. From the nature of the question itself it is clear that Herman sees mathematics as a precise subject in the sense that its conclusions are truths about the dice tossing process under consideration. He is surprised because he has painfully encountered a conclusion obtained by mathematical reasoning

(and interpreted in relative-frequency terms) that is false. Herman's faith in the precision of mathematical reasoning is misplaced. Mathematical reasoning is precise in the sense that its conclusions are valid with respect to the assumptions made; valid conclusions are not necessarily true statements about the process in question.

b. Herman's conclusion is correct only in the sense of being valid with respect to his probability model and the assumptions that it reflects; if we accept his probability model as a point of departure, then we must accept his conclusion as following from it.

c. What went wrong is that the probability model that came with Herman's dice was not a realistic reflection of the nature of his dice. Herman's probability model is a good fit to well-balanced dice, and he had a pair of "loaded" dice.

Section 10.1 (page 237)

1. Disagree. The parallel postulate problem was to either deduce Euclid's fifth postulate from his other postulates or replace it with an equivalent postulate which satisfied the criteria of being simple and self-evident.

2. Disagree. Lobachevsky sought to show that the system based on the postulate "If given a line L and point P not on L, there are at least two lines which pass through P and are parallel to L," taken as a replacement for Euclid's parallel postulate, together with Euclid's other postulates form a consistent system.

4. Disagree. Euclidean geometry is not synonymous with the term physical space. It is one possible model of physical space, so that if a system contains statements which contradict those of Euclidean geometry this does not mean that it stands in contradiction to the nature of physical space and therefore cannot be a realistic description of physical space.

5. Disagree. This statement confuses truth with validity.

6. Agree.

8. No; a postulate may be true, false or neither true nor false (all x's are y's, for example). It is a statement which with others form the foundation of a system from which valid conclusions are deduced.

9. Disagree. Gauss never undertook to establish such nonsense. Parallel lines, by definition, are lines in the same plane which have no point in common; there is nothing to prove here.

15. Disagree. More generally, it is possible to obtain true theorems in a postulate system which is based on false postulates. For example, consider the postulate system based on the following two geometric postulates: (1) All squares are triangles, (2) All triangles are rectangles. Theorem: All squares are rectangles.

18. Disagree. If we prove not-P a valid consequence of the postulates of geometry, then this is all that we have done. To prove a statement P in geometry by *reducio ad absurdum* we must show that the assumption of not-P together with the postulates of geometry leads to a contradiction.

20. The statement is, of course, nonsense, but one can feel sympathetic to its author after spending many hours in struggle with the subject.

21. Agree.

Section 11.1 (page 241)

2. a. No. The validity of the conclusions of Einstein's general theory of relativity has been established by mathematical reasoning (deductive-logical argumentation) and is therefore beyond dispute.

   b. The results obtained by the refined radar technique alluded to, and by any other experimental means, for that matter, pertain to the truth of the conclusions of Einstein's theory. The debate is over the realism of the conclusions of Einstein's theory, not their validity in the logical-deductive sense. The term "validity" as used in this quote is synonymous with realism.

3. The statement quoted means that a valid conclusion of the hypothesis of Einstein's theory of relativity was shown to be true by the experiment and observation means described. Whenever a fundamental valid conclusion of the hypothesis of a theory is shown to be true, this adds support to that hypothesis as a realistic description of the phenomenon in question.

8. a. No; the term "invalid" here should be replace by "unrealistic." Data obtained by experimental means is irrelevant to the validity issue in terms of its deductive-logical sense.

    b. As noted above, it is relevant to the issue of how realistic such theories are.

9. It's a question of what one means by confidence. If confidence is understood to mean that a theory of science contains the final truth on the matter in question, then the discovery that it does not can be most unsettling. A scientific theory, by its nature, cannot be the final word on the phenomenon under consideration. It is a model which, sooner or later, will require refinement as facts that deviate from its predictions (theorems) are uncovered. As facts come to light which agree with a theory's predictions, our confidence in the picture that the theory gives us is strengthened, but it would be a fundamental misunderstanding to view that picture as the final word.

    On this point it would be useful to review Questions 3, 10 and the comments noted on these questions.

10. a. No. The creation of a Bose-Einstein condensate has nothing to do with the validity of the predictions of quantum theory, which is a long settled mathematical (deductive-logical) issue.

    b. The creation of a Bose-Einstein condensate would show that one of the "outlandish predictions" of quantum theory is true. This would add support to quantum theory as a realistic description of nature and further increase our confidence in it being so.

14. The mathematical ideas inherent in a mathematical model may be ingenious, most sophisticated, and beautiful to behold as a product of the human mind, but if the model does not fit reality it becomes an imaginary universe which cannot be used as a portrait of reality.

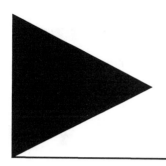

INDEX